Tracking the Bullet Saved My Child

Tracking the Bullet Saved My Child

A Compelling Story on How to Cut Down on Gun Violence

"Finally, a gun control law that works"

Demetri Bell

Copyright © 2010 by Demetri Bell.

Library of Congress Control Number: 2009912240
ISBN: Hardcover 978-1-4500-0071-0
Softcover 978-1-4500-0070-3

All rights reserved. No part of this book may be reproduced or transmitted in any form or by any means, electronic or mechanical, including photocopying, recording, or by any information storage and retrieval system, without permission in writing from the copyright owner.

This book was printed in the United States of America.

To order additional copies of this book, contact:
Xlibris Corporation
1-888-795-4274
www.Xlibris.com
Orders@Xlibris.com

61664

Contents

Dedication Page 7

Chapter 1   Why Are Guns Out of Control Today? 11

Chapter 2   Track the Bullet, Control the Gun 17

Chapter 3   A True Story 44

Chapter 4   How the Gun Regulation System Works 56

Chapter 5   Tracking Data to Help Solve Gun Crimes 66

Chapter 6   What Do the Gun Laws Show Today? 68

Chapter 7   Restrict and Enforce Versus Challenge and Prevent 125

I dedicate this book to the many parents who have experienced the fear that I have experienced while driving through the neighborhood streets of Chicago to pick up my son. When a young man pointed his hand in the form of a gun at my windshield and pulled the trigger, from a distance, it looked like a real gun and I ducked, slamming on the breaks of my car to try and avoid being hit by what could have been a bullet. The horror stories on the nine o'clock news of how many young people are being gunned down in the streets of Chicago in gun-related incidents perpetrated by young thugs really drove home the point that something needs to be done to put an end to this gun violence. Thankfully, it was not a real gun that this young thug was holding in his hand. However, I now fully realize what parents must be going through on a day-to-day basis, living under these conditions, where young adolescents are out of control. And they fear for their child's life as well as their own, not knowing when a stray bullet will claim their or a loved one's life.

Foreword

I wrote this book out of the growing concern I have for the number of young people who are being killed in this country from gun violence, often times by other young people.

Then I ask myself the question, just as so many other concern citizens and parents do, "Why aren't there any gun control laws to protect our children from gun violence by other children?"

Our kids are supposed to be our future; yet, we are letting them die because we have lost control of the way guns are being used in this country, often times by other young people who seek to destroy their future.

In this book I will reveal to you why the gun control laws that are in effect today can't solve the real problem at hand.

I will also reveal to you why a gun regulating system called "tracking the bullet" is the solution to controlling the way a person uses his/her gun and how it can largely reduce gun violence.

Once we do this, we would then start to save lives and give our children a chance to escape bullets that have claimed the lives of so many other young people.

Don't let your child become another statistic when you now have in your power to take back control of an uncontrolled situation.

When you have no gun control laws that address the real problem at hand, you have chaos, which is why you have situations where young teenage gang bangers are running around in the streets of Chicago killing at their own free will—killing students who could have grown up and found a remedy for cancer.

We would never know because their lives have been cut short at the pull of a trigger, shooting deadly bullets from the hands of an irresponsible one.

This book will appeal to parents, concerned citizens, and to those who are just tired of giving control to young adolescents and to those who use their guns irresponsibly.

Chapter 1

Why Are Guns Out of Control Today?

On July 2, 2008, Mike, who has the legal registration of a 45 automatic handgun, goes out to purchase a round of 9mm bullets for his gun, which cost him about thirty bucks. He then fills his gun with the bullets and places his gun inside of a shoebox in the closet on the shelf located behind his clothes.

He goes off to work. While he is at work, Stephen, his twelve-year-old nephew who lives with him, goes into the closet and finds Mike's gun and goes outside showing it off to his two friends, Jerome and Dwayne.

He then fires off several shots up into the air. Once he is done, the nephew puts the gun back into the shoebox inside of the closet, leaving the gun half empty. Two days later, on the Fourth of July, Mike wants to show his patriotism and celebrate the Fourth of July.

So he grabs his gun, goes outside with his friend, and shoots the remaining bullets up into the air, celebrating the Fourth of July. He then goes back to the gun shop to purchase more bullets.

Two blocks away, a little girl is found dead in her backyard with a 9mm bullet launched in her heart—young, twelve-year-old, Samantha, the daughter of David and Denise Williams.

According to the coroner's office, she had been dead approximately two to three days.

There she was, lying in the tall green grass that had not been cut for several days. The grass was so tall that had it not been for the pink dress that young Samantha was wearing at the time of her death, they could have easily missed locating her dead body so quickly. The pink dress stood out in the tall green grass where she lay.

The cause of death was identified as being hit by a bullet while she was sitting on her back porch, enjoying the nice summer breeze. When the bullet hit her, she obviously fell off the back porch, landing in the tall green grass in her backyard.

The police tell the parents, "This was probably caused by a drive-by shooting by rival gangbangers and a stray bullet hit and killed young Samantha during the process."

"In this neighborhood innocent kids are killed similar to this all the time," Officer Gary replies. The parents of young Samantha, David and Denise Williams, are outraged and they want a full investigation launched.

Area Detective Sharp and his partner Detective Jones are assigned to the case. They immediately start questioning the people in the neighborhood as to what rival gang did this or what happened here between July 2 and July 4.

No one else in the immediate neighborhood knew where, when, and why this happened—that a stray bullet came and claimed this little girl's life.

All they know is that a few days ago young Samantha was sitting on her back porch minding her own business, enjoying the breezy summer day, and now she's gone.

Detectives Sharp and Jones continue their investigation by contacting the neighborhood for information as to what gang-related drive-by shootings took place between the second and the Fourth of July within the surrounding area.

The informants tell the detectives that there was a drive-by shooting that took the lives of three youths, but this happened three blocks away from where the little girl lived.

The detectives drive to the area where the alleged drive-by shooting took place by rival gang members between July 2 and July 4, and they find 9mm bullet shells lying around on both sides of the street.

There were about fifty bullet shells found, which would corroborate the informant's story, but they cannot find the link between what happened here and three blocks away, where a stray bullet hit and killed young Samantha while sitting on her back porch.

Detective Sharp asks his partner, "How far does a bullet travel?" Jones replies, "About one to two miles. It takes eight blocks to make up a mile in this city."

"The victim lived just three blocks away, not even a half mile from where this shooting took place," replies Sharp.

Jones says, "Judging from the directions of these bullet shells, these shots were fired southwest and southeast. The victim lived northeast."

"The bullet would have to ricochet off something to travel in the northeast direction, dodging all of these buildings before it reached Samantha."

"And given the force from the ricochet hit and the distance it would have to travel to strike Samantha would have made it less lethal by the time it reached her."

Sharp says, "I see your point, maybe she had a boyfriend who was in a gang and he went over to her house and maybe they got into a lovers' quarrel and he shot her."

Jones replies, "That's one theory, but I still have not ruled out foul play here."

"Given where she was located at the time of her death, sitting on her back porch, would make it very difficult for a random stray bullet to hit her from this location."

"It looks more likely that the shooting was done by someone that she knew and in close range, opposite direction, but facing her."

"She also had to be standing, not sitting at the time that she was hit. Given the height of the back porch banisters, if she were sitting down, she would have fallen down on the porch floor up against the porch banister instead of falling off the porch over the banisters into the backyard grass."

Sharp says, "I see your point, let's take a ride over to the Williams' house and find out what type of friends Samantha hung around with."

They get into the unmarked squad car and head over to the Williams' house.

On their way over to the Williams' house, detective Sharp continues, "I wish we could track the source from where these young gangbangers are getting their guns, just like we did with that Tate bust last year, remember that?"

Jones says, "Yeah, who could forget it. That Tate guy was supplying all types of guns and ammunition to these young street thugs for a handsome fee.

Even if we could track the source from where these gangbangers are getting their ammunition, there will always be your Tate's arising on the scene. Someone will always be out there supplying guns illegally in this country to these street thugs.

What we need is a better system of stopping the supplier where they themselves cannot obtain the firearms to supply these street thugs."

Sharp says, "Even if you could stop the supplier from supplying these thugs, there are still enough guns out on the streets to last a lifetime. You don't really need a supplier of guns nowadays. Guns are easily obtainable without the supplier.

Even flea markets sell guns. What we need is a system where the supplier cannot purchase the ammunition to supply to these street thugs. Even if there are over a million guns out on the streets, with no bullets to shoot, what good are they?"

Jones says, "Yeah, you're right, but how would you come up with a system like this without violating someone's constitutional right to possess arms?"

Sharp says, "This would not be violating their right to possess arms, they will still have the right to possess arms, they just won't have any bullets in their pistols to shoot and kill anybody."

They both laugh at this as they arrive at the Williams' house. Detectives Sharp and Jones continue their investigation by meeting with the parents, David and Denise Williams, of the slain victim.

Denise asks the detectives, "Have you heard anything regarding the shooting of our daughter?" "Not yet, still investigating," replies detective Sharp.

Detective Jones asks Denise, "Tell me about your daughter, was she well liked by her peers?" Denise says, "Yes, very much so; my little angel had friends everywhere and she never did anything to hurt a soul. How could this happen?

We had such high hopes for her, she was an honor student in school; she loved math and wanted to be a doctor when she grew up so that she could help so many people who are constantly getting shot and injured by drive-by shootings.

She had a good heart. She was the type of child that even if you hurt her feelings, she would not show it. She would just be quiet and smile."

David replies, "We let her play in the backyard a lot with her friends due to the constant gang-related drive-by shootings."

Sharp says, "Did your daughter have any boyfriends that was in a gang?

Or what about the school she attended, did she ever tell you that she had any run-inns with gang members?" Denise says, "She attended Carter elementary on seventy-fifth and drake."

Jones says, "I have a daughter who attends there as well, maybe she knew your daughter, her name is Sabrina." Denise says, "That name doesn't ring a bell, and my daughter never mentioned having any run inns with gang members or any boyfriends that we knew of.

Now that you mention it, her brother had run inn's with some gang members, my son, Jason, he's thirteen years old."

Jones says, "May we speak with him?" Denise says, "He's not at home, he's spending the night over some relatives house. Can you wait until after the funeral is over?

He is very hurt over the loss of his sister." Jones says, "We understand, our sincere condolences for your loss."

Sharp says, "If you don't mind my asking, have the funeral arrangements been made yet?" Denise says, "Yes, It will be held at Peaks Funeral home on seventy-eighth and cottage grove this Saturday at 9:00 a.m."

Sharp is reluctant to ask this next question but he must cover all tracks during his investigation. Sharp asks, "Do you folks own a gun? David says, "I do." Sharp asks, "What type of gun do you own?"

David says, "A 45 automatic handgun. Sharp says, "May we see it?" David replies, "Yeah, be right back, it's downstairs in the basement." While David goes downstairs to the basement to retrieve his gun, his wife has tears coming out of her eyes, crying and feeling hurt over the loss of their daughter.

Meanwhile David brings the handgun up from the basement and hands it over to detective Sharp. Sharp investigates the gun and replies, "When was this gun last fired?"

David says, "About two days ago at the firing range. I go there often." Jones says, "Which location?" David replies, "The one located on forty-seventh and Cicero." Sharp says, "Can anyone verify that you were there on those dates?" David says, "Yeah, the owner can just ask for Pete." Denise says, "How do we save our thirteen-year-old Jason, who also may become a victim of gun violence with the increase of gun-related crimes in this city. What can we do as parents to protect our other child?

What can we do to stop this gun violence on our young people, I feel so helpless and useless, I need to do something so that nothing like this happens to some other parents' child as well as my other child."

Jones says, "Get together with other parents and concerned citizens in the community and push for tougher gun laws. It will not bring back your daughter, but at least by pushing for tougher gun laws, you can possibly prevent your son from becoming another victim once these tougher laws are put in place."

Sharp says, "Good day folks. We will be in touch if we get a break in the case." They now leave the Williams' home and head out to their car.

Detective Sharp tells detective Jones to check out David's alibi at the firing range on Cicero and forty-seventh and that they will meet back up tomorrow at the precinct and from there head over to young Samantha's funeral to pay their respect.

They will be on the lookout for any suspected gang members that might show up at the funeral to mock the ceremony. It is not uncommon for gangbangers to display their disrespect.

On the way home, detective Jones stopped by the gun shop on forty-seventh and Cicero to check out David's alibi, and it pans out. The owner of the gun shop confirms that David was there on the date and time mentioned to the detectives. Jones now heads home.

When detective Jones arrives at home, he immediately starts to question his daughter, Sabrina, to see if she knew the slain victim, Samantha.

Sabrina replies, "Yes, I knew her very well. She was one of my best friends, a very sweet person. Everyone liked her so much. The school is just not the same without her."

Jones says, "Did you know her brother as well." Sabrina says, "Yeah, Jason Williams, he's on the basketball team." Jones asks, "Does he seem like the type that would be in a gang?"

Sabrina says, "No way, he is way to cool for that and way too smart to be hanging out with such losers like Stephen, Jerome, and Dwayne."

Jones asks, "Are these other boys that you just mentioned in a gang?" Sabrina says, "Gang wannabees, the real gang members are the crooks and the Cobras.

The Cobras would be hanging out on the street corner around the school. Stephen, Jerome, and Dwayne would be over there, pitching pennies with them after school.

They want to feel like they are in the gang so much. They are always hanging around the Cobras, but they are nothing more than coward wannabees."

Jones asks, "Did Samantha have any boyfriends?" Sabrina says, "No, none that I knew of." Jones says, "Do me a favor, ask around at school to see if some of her friends that she hung around knew if Samantha had any boyfriends." Sabrina says, "OK dad."

Chapter 2

Track the Bullet, Control the Gun

The next day detectives Sharp and Jones arrive at the precinct and they head over to the funeral arrangement of the slain victim to pay their respect and to see if they notice any suspicious suspects like gang members.

No gang members show up at the funeral arrangement.

Hoping to make headway in the case, Jones says, "Denise mentioned that her son had run inn's with some gang members and my daughter mentioned three boys that hang out with gang members after school on the street corner. Is it possible that one of those boys had a run inn with Jason and then to retaliate they shot his sister?"

Sharp says, "Not likely, they would have shot Jason instead. Remember, these are kids; they don't want to send a message by killing the sister. They would much rather get revenge by killing the person that they had the confrontation with.

"Did you check out David's alibi?" Jones says, "I did, and it panned out. The firing range owner has a record of David being at the firing range at the date and time mentioned by David."

Sharp says, "I figured as much, besides there was no real motive for David to have shot and killed such a sweet daughter as his Samantha."

Two days after the funeral, angry parents and concerned citizens march in the streets of Chicago protesting gun violence, carrying signs that say stop the violence and asking for tougher gun laws.

The next day, community leaders, the mayor, and politicians meet with the parents and concerned citizens to discuss what can be done about the increase of gun violence, where young people are losing their lives.

The mayor explains that they have just increased police patrol in certain neighborhoods where gang violence is high. One angry parent shouts out, "Why can't you just ban all guns here in the city?"

The Mayor says, "Unfortunately I can't do that, this would be in violation of the second amendment of the constitution, a person's right to possess arms to protect themselves."

The Mayor continues, "We have gun control laws that have been passed, banning handguns. And if a person is caught using a handgun illegally, he goes to jail."

Denise says, "Mr. Mayor, with all due respect, even with the handgun banning laws in place today, they still do not address the real problem at hand. If they did, my dear Samantha would be alive today, but because of the irresponsible use of someone's gun, her life has been taken away.

Those gun-banning laws only address the consequences of what would happen for breaking the handgun banning law. That is of course, if the perpetrator is ever caught using or carrying handguns illegally.

This does not stop the perpetrator from committing the crime, we need laws that can prevent these young people from getting their hands on the guns and or the ammunition to fire deadly bullets that kill."

The Mayor says, "I understand your concern, how do you suggest that we do this? I am open to any and all ideas."

The room suddenly grew silent because no one could think of a way to prevent the young people from getting their hands on the guns or the ammunition.

And so the meeting was adjourned.

Meanwhile, the next day, young Jason is on his way to school; he is dribbling his basketball between his legs like basketball players do.

The young girls that also attend the same school are admiring him as he passes by since he is the star of the basketball team.

Sabrina who happens to be one of the girls admiring him asks her girlfriends, "Did Samantha have any boyfriends?" They all reply, "None that we knew of."

As Jason crosses the street, on the corner are Jerome, Stephen, and Dwayne pitching pennies with members of the Cobra gangbangers

Stephen yells out, "There goes that punk Jason dribbling his basketball in front of the girls like he's all that and a bag of chips."

Jason replies, "Yeah, I got your punk . . . punk."

Immediately the members of the Cobra gang say to Stephen, "You are going to let him disrespect you like that dude? You better bust a cap in that punk for disrespecting you like that."

Stephen replies, "I'm not worried about him, I can whip him anytime and any day."

The leader of the Cobra gang replies, "You either bust a cap in that punk for disrespecting you like that, otherwise don't bother hanging around us because we don't allow punks to be a part of the Cobras."

Shortly after that the school bell rings and Dwayne, Jerome, and Stephen head into the school for class. While in school, all Stephen can think about is what the leader of the Cobra gang told him about busting a cap in Jason.

The school bell rings, it's now recess time. During recess time, all the other kids are outside playing in the playground. Stephen, Jerome, and Dwayne are sitting on the school steps, plotting to shoot young basketball star, Jason.

Dwayne says to Stephen, "How are you going to handle this dude?"

Stephen asks, "Handle what?"

Jerome says, "You know good and well what."

Stephen replies, "Where would I get a gun from to bust a cap in that punk Jason?"

Dwayne says, "Hey, what about that gun that you showed off to us and fired several shots up into the air?"

Stephen says, "That was my uncle's gun."

Jerome says, "So, he won't know that you used it, just do like you did the last time and put it back where you got it from."

Stephen is thinking to himself that this could possibly work. The school bell rings, signaling that recess time is over and time to return to class. They disburse and return to class.

While in class, the only thing that Stephen can think about was how the members of the Cobra gang taunted him about letting Jason disrespect him, and he would not be accepted by the Cobra gangbangers if he did not bust a cap in Jason.

The school principle announces over the PA sound announcement system that four weeks from today there will be a basketball game between Carter elementary and Powel elementary on a Tuesday, from 5:00 p.m. to 7:00 p.m.

He tells the students to come out and support the Carter wildcats as Jason and the Carter wildcats take on the Powel warriors.

Tickets cost $5 first-come, first-served basis. When Stephen hears Jason's name mentioned over the PA sound system, he grows angrier and angrier by the minute as the plot to shoot Jason thickens. The time is now 3:00 p.m.

The school bell rings, signaling that school is over. The students head out to go home. Jerome, Stephen, and Dwayne go to their usual hang out place across the street on the corner to pitch pennies with members of the Cobra gangbangers.

The Cobras waste no time in taunting Stephen about letting himself be disrespected by Jason. They said that he just stood there like a little punk and did nothing.

By this time Stephen was so embarrassed and angry that he leaves the penny pitching ceremony earlier than usual because he can't take it any longer, their poking fun at him and calling him a punk that got disrespected.

Jerome and Dwayne just look on as Stephen walks away to go home, but they dare not leave and follow him home for fear that the members of the Cobra gangbangers might call them punks as well for leaving early with Stephen.

Meanwhile detectives Sharp and Jones are at the precinct discussing gun control laws.

Sharp says to Jones, "Remember that last discussion that we had about gun control and I mentioned to you that if we could come up with a way to stop these young people from getting their hands on the bullets, then we can control the way they use their guns." "Yeah I remember," replies Jones.

Sharp says, "What if we came up with a system where we can track each bullet fired from a person's gun. A computer system that tracks every bullet that a person purchases and then makes them accountable for the way they use those bullets, thereby controlling the way they use their guns."

Jones says, "How would we do this Einstein?"

Sharp says, "By making a person accountable for every bullet that he purchases every single time. Even if it is just one time where he fails to account for any bullet that he purchased and has not used it in a responsible manner, then he is deemed to be using the bullets irresponsibly to kill or to injure. He will be banned from purchasing bullets.

Through the process of elimination, you then root out the source that is supplying these young gangbangers with the bullets that they shoot from their guns to kill people.

People who are purchasing bullets and irresponsibly selling those bullets or supplying these bullets to the street thugs will be exposed and banned from purchasing bullets since this is an act of irresponsibly using bullets that kill.

Those who purchase bullets and fire those bullets from their gun irresponsibly will also be exposed and banned from purchasing bullets."

Jones says, "What would be considered responsibly using their bullets?"

Sharp says, "Going to the firing range for target practice or to a hunting site to hunt animals is considered responsibly using their bullets, not purchasing bullets and then illegally selling or giving them to someone who then uses the bullets to commit gun homicides.

This is an act of irresponsibility, using bullets that you purchased by giving the bullets away to someone who then commit a gun crime.

A person who randomly shoots his bullets up into the air without accountability as to where those bullets will end up is also considered irresponsibly using bullets that kill.

Those bullets could have came down and hit and injured or killed someone. Since these are acts of irresponsibly using bullets that kill, those persons would be banned from purchasing bullets for their irresponsible acts.

The computer system will track a person's bullet usage against the amount of bullets that he purchased verifying that he used all of those bullets at the firing range or at a hunting site.

If one bullet is missing or goes unaccounted for, he is banned from purchasing bullets, because that one bullet that went unaccounted for could have been used to commit a gun homicide."

Sharp continues, "With the technology that we have out here on the market today, it would not be hard to put something like this together. The gun-control system would be called "tracking the bullet to save lives."

The way it would work is that everyone that has the legal right to purchase bullets would receive an electronic swipe card in the mail with a personal identification number or PIN for that swipe card."

Jones asks, "A swipe card?"

Sharp says, "You know like your bank debit card, when you go to your bank to swipe your card through the cash station machine, it asks you for your personal identification number or PIN.

You then enter your pin and the banking system checks your data to make sure that your credentials match.

Every bank debit card has a computer database associated with that account. The technology would be similar to this.

Every gun shop and firing range in America would have a computerized cash register with an attached swipe card reader for reading your swipe card and a electronic keypad for entering your personal identification number.

That data would then be transmitted to a central database on the computer located at the local police station in every neighborhood in the country.

The firing range, gun shops, hunting sites, and the police station will all share the same database information for accuracy of tracking the bullets that people purchase for their guns per individual swipe card personal identification number or PIN.

If an individual does not have a swipe card, then they cannot purchase bullets; likewise, if a person forgets his personal identification number, a new one would have to be issued to them before they could purchase bullets again. Until then, they cannot purchase bullets without it.

There should be metal detectors installed at every firing range in the entrance to the firing range and a clerk stationed next to it.

Also, metal detectors should be installed in the entrance to the gun retail shop that is within the firing range."

Jones asks, "Metal detectors?"

Sharp says, "The purpose of the metal detectors is to check if a person has bullets hidden on their person, in their pockets, when they entered into the firing range and when they leave, so that every bullet can be tracked and accounted for.

The clerk will enter the total number of bullets found into the computer system and give the customer a receipt showing how many bullets the clerk found in their gun and on their person.

To keep up with and track the number of bullets per individual swipe card associated with the number of bullets that was purchased by that individual, this same process is repeated when the customer leaves."

Jones says, "Oh I see, a person goes into the firing range, his gun is checked for the number of bullets that he has in his gun before he fires shots from his gun, and it is also checked when he leaves for accuracy of tracking the bullets that he used at the firing range.

He is also scanned by the metal detectors for bullets found on his person for accuracy of tracking and the total number found will be entered into the computer system database. And this is tracked against the total number of bullets that he previously purchased for his gun associated with his swipe card personal identification number.

The computer would track this after the shop clerk enters the number of bullets found in the person's gun and on his person when entering and leaving against what the customer previously purchased."

Sharp says, "Yes, that is correct. The next time he goes to purchase bullets, the computer would check the number the customer previously purchased and what was actually used at the firing range based on the data input into the computer by the shop clerk.

The number used must match the number previously purchased for accuracy of tracking every bullet that he previously purchased and used.

If one bullet goes unaccounted for, he is denied access to purchasing bullets until he can responsibly account for all the bullets that he purchased and used.

The computer software is designed to ask predetermined questions that would determine if the customer used those bullets responsibly or not.

The clerk will enter the answers to those questions into the computer system from a certifying application that the customer must fill out to qualify to repurchase bullets again.

The application is called a recertifying application, certifying that you used your bullets responsibly. The store clerk will hand this application to the customer to fill out at time of repurchase.

The answers to those questions that the customer provides on the recertifying application will show if the customer used those previous bullets responsibly or not."

Jones asks, "What about a person who makes his own bullets. How will the tracking the bullet gun law track his bullet usage?"

Sharp says, "The same way that the tracking the bullet database tracks a person who purchases whole bullets.

Remember, in order to make your own bullet, you must have bullet shell casings and gunpowder and bullet heads. The powder is placed into the bullet heads. And the bullet heads are place into the gun shell casing. When shots are fired, the gun shell casings fall to the ground and maybe kept for reuse.

When he goes to buy the bullet heads from the gun shop, this will be tracked the same way whole bullets are tracked by the Tracking the Bullet database system."

Jones asks, "What about a hunter who uses his gun at a hunting site and not at a firing range. How will his bullet usage be tracked for responsible usage?"

Sharp replies, "The rule that applies for the hunter at a hunting site is the same as that applies for the person going for target practice at a firing range.

There will be a clerk at the hunting site who will record the number of bullets in the hunter's gun before he hunts. The clerk will also check to see how many bullets he has in his gun before he leaves and when he exits the hunting site.

The computer will track the number of bullets that he purchased against the number that he used at the hunting site.

Again the number purchased must match the number used at the hunting site. If there is a mismatch, then he is banned from purchasing bullets until he can responsibly account for the mismatch.

There should also be metal detectors installed at the hunting site location where the hunter goes to check in so that he may be scanned for any bullets that may be hidden on his person or in bags. Again, this is for accuracy of tracking the bullet.

This should be done before they start hunting at the hunting site."

Jones says, "Instead of using metal detectors why not just have the customers empty their pockets and bags that they may have with them for any bullets that they may be hiding on their person? This would be a lot cheaper than having metal detectors installed, don't you think?"

Sharp says, "Yeah, that could work, but for accuracy of checking I would go with the metal detectors.

This system is designed to prevent an individual from hiding bullets on his person and then telling the site location clerk that he actually did use those bullets to hunt. And then the clerk enters that number into the tracking system. Later on, the person uses those hidden bullets to commit a gun homicide."

Jones says, "What about the hunter who lives in rural areas where there is no computerized hunting site for hundreds of miles from where they live? And so their land is the only place that they can legally hunt deer on. And there is no real statistical gun crimes mentioned in that location where they live. A good example of this would be the state of Montana."

Sharp says, "There may be some special allowances under the Tracking the Bullet gun law for those hunters who live in rural areas where there is no computerized hunting site for hundreds of miles, and no real gun violence is happening in those locations such as the state of Montana. Farmers would also be a good example of this.

Then in cases like this, a lower degree of accountability could be imposed under the Tracking the Bullet gun law for situations like these.

If the location where he is hunting is his own personal land and there are no gun-related homicides known in that location, then a lower degree of accountability maybe imposed for this unique purpose. I mean, it's only fair.

As long as the hunter can show that those bullets will be used on his personal land and the location where he lives at has little to no gun homicide rates statistically speaking, then a lower degree of accountability maybe imposed.

However, if the gun crime homicide rate starts to go up in that location, then a higher degree of accountability would be imposed under the Tracking the Bullet gun control law.

Again, the exceptions are that the location must be an area where there are no gun-related homicides happening and the hunter must use his bullets on his own land to hunt animals.

However, the whole purpose of the Tracking the Bullet gun law is to save lives, not to convenience people."

Jones says, "I agree that there should be degrees of accountability imposed within the Tracking the Bullet gun law—certain amendments and exceptions depending on location where hunting activity will take place and whether there is no real gun homicides happening there and it is being done on their own land."

Sharp says, "The hunter must also provide a valid copy of his hunting license, it must not be expired. He must also explain on the certifying application how many bullets he used at the hunting site and what type of bullets he used at the hunting site again for accuracy of tracking those bullets.

By making a person accountable for the bullets that he purchased and used, think about how many lives we can save by implementing this.

No more can a person just go out there and repeatedly use his gun to commit gun homicides with deadly bullets and not be caught for long periods of time and in a lot of cases, never being caught at all. Now his bullet usage is being tracked and he will be stopped dead in his tracks after his first incident of irresponsible bullet usage.

Or maybe he is a person supplying these young street thugs with deadly bullets that kill. This will come to an end under the Tracking the Bullet gun law because now his bullet usage is being tracked.

If he did not use those bullets at the firing range or to defend himself legally under the second amendment or for hunting purposes or at a hunting site, then everything else is considered irresponsibly using bullets that kill.

This person will then be banned from purchasing bullets. His swipe card personal identification number will be deactivated by the computer database system.

Granted the Tracking the Bullet system will not be a cure all for every gun-related homicide that happens out there, but it will put a huge dent in the amount of irresponsible, reckless usage of bullets by these young street thugs and gangbangers in major metropolitan areas, which is claiming innocent lives.

They are not hunters nor do they go to the firing range.

They are killing innocent people at their own free will and some of them never get caught or have to give an account for what they are doing with their guns.

And so the current gun control laws can't stop them.

Our children's lives are being taken on the streets of Chicago, Detroit, New York, Los Angeles, and other major metropolitan areas by the irresponsible usage of someone's gun.

This is where the bulk of the gun homicides are happening."

Jones says, "I see, so a person would have to recertify that he is using his bullets responsibly every time he goes to repurchase bullets?"

Sharp says, "Yes, he is held accountable every time he purchases bullets that kill."

Jones says, "I am curious, what would be some of the questions on the certifying application?"

Sharp says, "The first three questions on the certifying application are the most important, which would show immediately if a person has used his bullets responsibly or not.

There are only two possible options that a person can choose from. The choice that a person makes will determine if they can purchase bullets right away or have to wait a few days until their proof of acceptable documentation is checked out and verified that they have used their bullets responsibly."

Jones asks, "So what are the three questions?"

Sharp replies, "Question number one would be, 'How have you used your bullets that you last purchased?'" And the only two options would be 'For target practice or Other.'"

"Question number two would be, 'Where did you use your bullets?'"

"And the only two options would be 'At the firing range or Other.'"

"Question number three would be, 'When did you use your bullets, date and time?'"

"If a person picked option one for the first two questions, which is target practice and firing range, then the computer database would have a tracking record of this. That is, if they are telling the truth. The same holds true for the hunting site.

The computer system would have tracked their bullet usage at the firing range since it shares the same computer database of the centrally located computers at the gun shops and the police stations and data is replicated throughout the databases for accuracy of record keeping.

They are allowed to purchase bullets right away, since they used their bullets responsibly and this was accounted for and tracked by the computer database system.

If they choose the 'other option,' then they would have to furnish acceptable supporting documentation showing that they used their bullets responsibly."

Jones asks, "How would a person choose the 'other option' and prove that he used his bullets responsibly?"

Sharp says, "Let's say a person broke into your house and attacked you and your family and you used your gun to defend yourself under the second amendment right of the constitution thereby shooting the perpetrator in the process of protecting you and your family."

When you choose the 'other option,' you would then state this on the line and supply a police report or other acceptable documentation verifying your story."

Jones asks, "What if the police report does not say that this is what happened?"

Sharp says, "Then you would have to supply court records such as a copy of the transcript vindicating you at your court trial hearing that this is what took place and that you did use your bullets in self-defense.

When you submit the acceptable documentation, this gets mailed in along with your certifying application to the local police station to review.

Then specially trained hired personnel would review the documentation, and they would determine based on your documentation of proof if you used your bullets responsibly or not.

They will be basing their decision as to whether you used your bullets responsibly in harmony with the second amendment of the constitution, your right to possess arms to protect yourself.

This is not a decision that the store clerk can make.

If you submit a copy of the transcript court records that vindicated you at your court hearing, this is all they would need to see in order to approve you repurchasing bullets.

Or, if a copy of the police report states this as well, this would also be sufficient."

Jones says, "Good job, Watson." Sharp says, "Thank you Sherlock."

"Now let's head over to the Williams' house to see if they can organize a meeting with the state law makers and push to get this put into the state gun regulating protection laws.

Then push to help make this a federal law. This gun regulation law would certainly help prevent these young thugs from getting their hands on bullets that kill and start to save lives in the process."

They arrive at the Williams house and they mention the plan to Denise and ask her to organize a meeting with the state law makers to discuss pushing this to make it not only state law, but also eventually federal law.

Denise is excited about the plan and thanks the detectives for their efforts and ideals to make the community a safer place to live not only for the young people, but for the adults as well.

Denise immediately meets with the community leaders to set up a meeting with the state congressman and mentions the plan to the state law makers with hopes of making this gun regulation a state law and then eventually a gun-regulating federal law.

The meeting arrangement is accomplished and the State congressman asks to see a copy of the plan and he is impressed with what he sees and hears.

He likes the idea, especially since it does not violate a person's second amendment right. He also sees where this plan works more toward prevention. He is also pleased with the fact that there are varying degrees of accountability depending on circumstances under the tracking the bullet gun law. He is also impressed how this will create more jobs and stimulate the economy. And he also sees how this can in fact save lives by rooting out most of the local irresponsible usage of a person's gun.

By keeping these deadly bullets out of the hands of these young gangbangers. Thereby rendering their gun usage obsolete.

He then explains to them what needs to be done to make this a state law and eventually a gun-regulating federal law.

He explains that he will show this plan to his constituents to make it a bill, allow the bill to be given a "second reading" and accomplish a favorable verdict by getting as many representatives on their side as possible. This may be achieved by generating public support and pressure.

Then the bill will be passed around to be given a third reading, and then allowed to be passed to Congress, who will then decide whether it is worth pursuing or not.

If so, it will then be passed to the president. Assuming that the president votes on the bill, and if he agrees, the bill becomes a law.

When the president signs the bill, the federal law will have been passed. If he vetoes it, they may still be able to get the law passed, provided the majority of the house votes in favor.

The whole process took just three weeks, and the bill was approved by the law makers and the president and it is now a gun-regulating federal law.

A letter from the federal government went out in the mail to all the gun registrants and people who have a legal fire arms identification card to purchase bullets and guns.

Mike, the owner of a 45 pistol automatic goes to his mailbox and retrieves the letter from the federal government concerning the new gun regulation law that went into effect.

The letter reads "Dear Michael Banks, Effective immediately in lieu of the new gun-regulation law titled "Tracking the Bullet to Save Lives" under section code 23456, article 5, you are required to give an account for each and every bullet that you purchase. You will be receiving a fire arms identification swipe card in the mail shortly.

This will be used every time you purchase bullets for the purpose of tracking each bullet that you purchase.

The first time that you use the new system you will not be required to certify that you have used previous bullets responsibly.

Thereafter, you will have to recertify every time you want to repurchase bullets for your gun. Certifying that you used your bullets responsibly.

If you cannot account for each and every one of the previous bullets that you purchased, you will be denied purchasing bullets until you can verify that you used every bullet responsibly. There are no exceptions to this rule.

The whole purpose for this law is to crack down on illegal irresponsible usage of the many guns that are out there on the streets claiming innocent lives."

Two days later Mike receives his electronic FOID swipe card in the mail. And he realizes that his gun has no more bullets in it.

So Mike goes to the gun shop to purchase bullets for his gun. The store clerk tells Mike to swipe his FOID card through the swipe card reader attached to the computerized cash register.

Mike swipes his electronic card and enters his personal identification number, his record appears on the screen of the computerized cash register showing that this is the first time he is purchasing bullets since the new gun regulation law went into effect.

The clerk then enters the number of the bullets that Mike is purchasing into the computer. Mike is purchasing six bullets for his gun.

The clerk then gives Mike two receipts, one showing the price and the taxes that he paid for his bullets and the other showing him the number of bullets that he purchased.

The clerk tells Mike to keep this for his records to verify that he purchased six bullets and that the next time he comes in to purchase bullets, he will have to certify that he used those previous bullets responsibly. Mike then loads the bullets into his gun and returns home and places his loaded gun in his usual hiding place—inside a shoebox in the closet on the shelf, behind his clothes. He then heads out to work.

Meanwhile, his nephew, Stephen, comes home from school. His two friends, Jerome and Dwayne, come over to hang out with him.

They start plotting about how the shooting of young basketball star Jason should be carried out. Jerome says to Stephen, "Have you ever aimed a gun and shot at a target and hit it before?"

Stephen, who does not want to seem like he has no experience aiming and hitting a target replies, "Yeah dude."

Dwayne says to Stephen, "OK, let's see because it's not as easy as you think. Let's go out back in the alley and set up a target and see if you can hit that target about eight feet away."

So they head outside in the back alley and turn over a garbage can upside down and they place an empty beer can on top of it.

Dwayne says, "See if you can hit this beer can." Stephen goes into the house where his uncle keeps his loaded gun in the closet and he retrieves it.

He returns and he stands eight feet away from the target and he fires one shot at the target and he misses it completely.

Neighbors hear the shot and someone raised the window to see where that noise came from. The three boys dart back into the house so that they are not seen by any of the neighbors.

Stephen puts the loaded gun back into the shoebox inside his uncle's closet on the shelf, behind his uncle's clothes.

The three boys leave out to go outside in the front and they meet up with members of the Cobra gangbangers.

It has now been four weeks since the announcement of the basketball game between the Carter wildcats and the Powel warriors. Two more days to go before the basketball game.

The leader of the Cobra gangbangers, whose name is Tyrone, says to Stephen, "Next Tuesday is your chance to bust a cap in that punk that disrespected you. Are you ready?" Stephen replies, "Yeah man."

Tyrone says, "You can do it after the basketball game is over. It will be dark in the evening and you can wait until he comes outside and starts walking home, then you can do it. We will do a drive-by and you bust a cap in him."

Stephen asks, "What if someone sees me and identifies me after the act and I go to jail for shooting and possibly killing Jason?"

"Can I just give him a beat down?" I did this when I was much younger, in the second grade."

Tyrone says, "Fool you aren't in the second grade no more. Now, do you want to be a member of the Cobra gangbangers or not?

"Wasn't it you that told me a few weeks ago that you are ready to be one of the boys and be cool and tough and have a lot of girlfriends because of who you are?

Don't you want to belong to a gang that would take care of you and never leave you and would be there for you whenever you need us?

And if you are worried about being seen, don't worry about that because you will be wearing a ski mask so nobody will be able to identify you.

The only thing that would be showing is your eyes.

I have a gun that you can use to carry this out because I know you don't own one. Here it is, but I do not have any bullets for the gun right now.

My supplier is supposed to be supplying me with some bullets for the gun in a couple of days.

You and I can meet up then and I will give you a few bullets for the gun, but make sure that you don't miss because bullets are becoming more difficult to get these days."

One of the Cobra gangbangers whose name is Little Dan, Tyrone's right hand man, replies "What's up with that dude, we use to be able to get plenty of bullets for our guns. Now the word out on the streets is that suppliers are having a difficult time getting us the bullets."

Tyrone says, "Fool, don't you watch the 9:00 p.m. news? Haven't you heard about the new gun regulation law that was passed called Tracking the Bullet?

Our suppliers now have to give an account for every bullet that they purchase since this is being tracked by a computer database system.

If they cannot prove that they used those bullets responsibly, then they will be denied access to purchasing bullets by the tracking software database system.

I know this one supplier who may have a few bullets left before they passed this new gun regulation law and he is going to try and get me some of those bullets in a couple of days."

Little Dan replies, "How are we going to defend ourselves against rival gangbangers if we do not have bullets for our guns?"

Tyrone says, "I am sure that the rival gangbangers are asking that same question. Looks like we will be going back to the way they did it in the old days when they fought with their fist."

Meanwhile Stephen accepts the gun from Tyrone and hides it and takes it upstairs and puts the gun up into his bedroom hidden away in his bedroom closet.

The next day Stephen, Dwayne, and Jerome are just walking, talking and hanging out while they discuss the plan to shoot Jason next week, Tuesday, after the basketball game.

Stephen now reveals his true feelings about shooting Jason to his two buddies.

Stephen says, "I remember when me and Jason use to play such games as hide and go seek, and Johnny come across when we were in the second grade, we had so much fun together.

Jason was my best friend back then, but as we started getting older, Jason started hanging around people who were more into sports like basketball and football.

And I started hanging around you two losers."

Dwayne and Jerome both laugh at this and say, "So what happened to the friendship?"

Stephen says, "One day Jason was with a couple of his basketball buddies and I was throwing little pieces of pebble rocks at the birds and they started making fun of me saying that I was an idiot for throwing at the birds like they were better than me or something."

Dwayne says, "Jason too?" Stephen says, "Especially Jason, they were laughing at me calling me the neighborhood bird bully and from that day forward we just stopped talking to each other.

I really never disliked him to the point of wanting to shoot him though.

I realize that he was just showing off in front of his new friends, but even if that's all he was doing, I still want to become a Cobra gangbanger. So I guess I am going to have to get use to the fact of busting a cap in punks like Jason."

Jerome says, "Yeah dude, now that's what I'm talking about. Just think about it, once you become a Cobra gangbanger and word gets around.

Nobody would be disrespecting you anymore.

You could go up to the most popular guy in the school and take his girlfriend and have him give you his lunch money every day because of who you are."

Dwayne says, "Now that's what I call cool."

Jerome says, "Wasn't there another time last year that we had a run in with that punk Jason?"

Stephen says, "Oh yeah, that's right. We were walking through the school hallway and Jason came running out of the gym room playing and being chased by beautiful Sabrina when he bumped into Jerome."

Jason said, "Excuse me, my bad, I did not see you." Then Jerome said, "Well if you would turn around and watch where you are going maybe you would see me punk"

Then Jason replied, "Whatever man." Then all three of us held up are gang symbols and yelled out loud "Cobra love."

Dwayne says, "Yeah, and did you notice that disgusting look that Sabrina gave us?"

Jerome says, "That was because she likes that punk Jason"

The three boys now head over to their usual corner to hang out with members of the Cobra gangbangers to pitch pennies with them.

Meanwhile Stephen's uncle, Mike, arrives home from work. He gets on the phone to call up his buddy, Ricky, to go over to the shooting range with him for some target practice.

Mike drives over to Ricky's house to pick him up as the two head over to the shooting range. Once they arrive at the shooting range, they enter inside the range and a clerk standing next to a metal detector and a laptop computer connected to a computerized network. The clerk greets them

The clerk tells them to pass through the metal detector to see if any bullets are found on their person. The clerk next checks their pistols to see how many bullets are in their guns. The clerk now enters the total amount of bullets found into the computer system. She found five bullets in Mike's 45 pistol automatic, and she hands him a receipt showing that this is the number entered into the computer system for tracking purposes.

Mike realizes that he previously purchased six bullets for his gun and cannot understand why there were only five bullets in his gun.

What he doesn't know is that his nephew, Stephen, fired one of those bullets from his gun in the alley back at home while practicing trying to hit a target for the purpose of shooting Jason.

Mike and Ricky then proceed to the firing range. Mike shoots all the remaining five bullets from his gun.

He then heads over to the retail shop within the firing range to purchase more bullets for his gun.

As he enters the retail shop, another clerk standing next to a metal detector and a laptop computer hooked up to a computerized network greets him.

He passes through the metal detector, the clerk now checks his gun for ammunition and enters zero amount of bullets found on his person and in his gun into the computer system.

The clerk now directs Mike up to the counter to purchase more bullets for his gun. Mike goes up to the counter and swipes his card through the electronic card reader and then he enters his personal identification number using the electronic keypad.

The computer's cash register displays his information on the computer screen with a message stating that all bullets have not been accounted for since his last purchase. One bullet left unaccounted for since his last purchase.

The clerk then explains to Mike that he purchased six bullets previously but he only used five bullets according to the computer tracking mechanism.

The clerk tells Mike that he has to account for that last bullet before he can repurchase bullets again. Mike explains to the clerk that he does not know what happened to that last bullet and that he thinks that he may have lost it.

The clerk explains to Mike that the whole purpose for the new gun regulation law called "Tracking the Bullet to Save Lives" is to make sure that a person is using the bullets that he purchased responsibly and that losing his bullet is not an option because the Tracking the Bullet System holds a person responsible for every deadly bullet that they purchase.

The clerk hands him a recertifying application to fill out and tells him that he can try and put on the application that he lost one of his bullets on the other option line on the certifying application.

However, the clerk does not believe that the system will accept this. Mike then fills out the application and fills in the other option field that he lost or misplaced his sixth bullet.

The system rejects this excuse and denies him purchasing bullets until he can responsibly account for that sixth bullet.

Mike is frustrated. The clerk explains to him that someone could find that bullet that he lost or misplaced and go out and murder someone with that bullet.

Although Mike was not the one that committed the murder, it was neglect on his part for failing to keep track of all of the deadly bullets that he purchased.

So that this does not happen again as a repeated irresponsible act on Mike's part, the system locks him out from purchasing bullets for his gun as a safety to the public.

The clerk further explains that there is a ninety-day waiting period before Mike can write a letter to the local law enforcement agency explaining that he honestly lost his bullet and that he has no prior criminal activity background and would like to be reactivated into purchasing deadly bullets again.

The clerk continues, "You may be reactivated at the discretion of the law enforcement agency. However, if this happens again, you will be forever restricted from purchasing bullets again due to your neglect."

Mike responds, "Is there a guarantee that I will be reactivated?" The clerk responds, "No, because these are peoples' lives that we are talking about here.

Someone can be seriously hurt or killed with just one bullet that you failed to keep track of. So the answer to your question is No.

However, the law enforcement agency will look at your tracking history as one of the determining factors in their decision to reactivate you or not. They will look to see if you have a history of using your bullets responsibly and all deadly bullets were responsibly accounted for, such as at the firing range or hunting site.

This will help them to determine if this could have been an actual mistake that you could have very well lost your bullet.

However, since you have no tracking history that shows you have used your bullets responsibly simply because the tracking the bullet system just started and your first tracking history shows you to be an irresponsible person, a person that can't keep up with his deadly bullets, it is more than likely you will not be reactivated."

Mike understands and leaves without being able to purchase bullets for his gun since he can't account for all of his bullet usage.

He still cannot figure out what happened to that sixth bullet. He would never think in his wildest dream that his twelve-year-old nephew would have taken his gun and actually shot it. Mike heads home and puts his empty gun back into the original hiding place in his closet.

Meanwhile Stephen, Jerome, and Dwayne are still on the corner pitching pennies, Tyrone pulls up in a car and yells across the street to Stephen to come on over to the car.

Stephen comes over to the car and Tyrone hands him five bullets and tells Stephen, "I paid a lot for these bullets. They have become harder and harder to obtain since the new gun regulation law went into effect, so make it count dude and bust a cap in that punk for disrespecting you like that.

And don't miss. My supplier had to scrape up some previous bullets that he had hidden in his house before the new gun regulation law was passed and he charged me a pretty penny for these bullets.

You can pay me back later for the money that I paid for these bullets by selling some rocks on the streets for me to make up for what I spent on these bullets.

This is your chance to prove that you are worthy to become one of the Cobra gangbangers. This is your initiation.

If you actually bust a cap in that punk, then you will become a Cobra gangbanger from that day forward with all the perks since you have proven yourself.

So don't let me down dude." Stephen nods his head signaling yes, and after accepting the five bullets he puts them in his pocket and returns across the street to his penny pitching ceremony.

Stephen, Jerome, and Dwayne decide to leave and return home. As they are leaving, one of the Cobra gangbangers hands Stephen a plastic bag with a ski mask inside the bag to hide his face while he carries out the shooting.

The next day is Monday morning and Stephen, Jerome, and Dwayne meet up to head out to go to school together.

All Stephen can think about is building up enough nerve and guts to actually shoot Jason and possibly kill him in the process.

While they are walking to school, Dwayne and Jerome notice that Stephen is very quiet. And they ask him, "What's up man, why are you so quiet?"

"Tomorrow is your big day to prove that you are worthy to become a Cobra gangbanger."

Stephen can barely hear them while they are talking because he is thinking about when Jason and he were younger, the smile on Jason's face when Jason tagged him while they were playing the Johnny come across schoolyard game after school. And now he is picturing the look on Jason's face when he shoots him.

At this point he is feeling guilty about shooting his former friend, Jason. However, peer pressure and his own selfish longing for wanting to be a part of the Cobra gangbangers overrides his guilt.

And so he now starts to focus on just how he will carry this out.

As the three boys arrive at school, they agree to meet up on the school steps during recess time to discuss how the shooting will be carried out.

While in school, the principal announces over the PA sound system that tomorrow, Tuesday, is the big game between the Carter Wildcats and the Powell Warriors. So come on out to support your team as Jason and the Carter Wildcats take on the Powel Warriors.

There are still a few more tickets left. Stephen is now thinking about how will he be able to shoot Jason after the game when there will be people surrounding Jason.

He is also concerned about missing his target and hitting someone else.

The school bell rings, signaling that it is now recess time and the kids go outside to play in the school playground.

Stephen, Dwayne, and Jerome meet on the school steps as the plot to shoot Jason thickens.

Stephen says, "What if someone pulls my ski mask off my face after I shoot Jason to identify who I am?"

Dwayne replies, "Dude believe me when I tell you that nobody is going to do that, everybody is going to be too busy running, trying to get out of the way."

Jerome says, "Not only will they not do this, but also they will not be able to do this since you are going to be inside of a moving car."

Dwayne says, "That's right, this is a drive-by shooting, so you better aim straight dude."

Jerome says, "If he aims like the way he aimed during that target practice back at the alley, then he is going to miss hitting Jason completely."

Dwayne says, "That's right, he might even hit someone else in the process.

Are you sure you want to go through with this Stephen? It's not too late to back out.

Just tell Tyrone that you can't shoot straight and that you don't want to hit the wrong person."

Jerome says, "Fool, you better not tell Tyrone that because he is going to call you a punk for being concerned about hitting the wrong person.

This is war and a Cobra gangbanger has to be tough, they are not concerned about hitting the wrong person like a little girl."

Stephen thinks to himself that Jerome is right and he cannot back out now, he must go through with this.

Stephen says, "I have five bullets. I am bound to hit him at least one time out of the five tries."

The school bell rings, signaling that it is time to return to class. As Stephen and Jerome and Dwayne head back to class, Jason passes by and looks straight into the eyes of Stephen and with a friendly smile says to Stephen, "Are you coming to the game tomorrow man?"

Stephen looks surprised that Jason is speaking to him in a friendly tone because he recalls the last conversation between the two was not so friendly.

This was Stephen's perception of the last conversation between the two.

Stephen for a brief moment is feeling the same feelings for Jason that he felt when they were friends together in the second grade.

Stephen just nods his head, signaling no to Jason and walks on.

As Stephen returns to class, he is thinking about the friendly smile and the nice tone, the way Jason spoke to him, and now he is wondering whether Jason actually disrespected him when he said "I got your punk . . . punk."

This only makes matters more complicated for Stephen in his decision to carry out the shooting of Jason. He is now thinking that he may have blown this whole thing out of proportion or the peer pressure from the Cobra gangbangers is the only reason why he is carrying this out.

However, he can't help but feel what it would be like to be a Cobra gangbanger with all the perks. So he psyches himself into believing that Jason still disrespected him and that he must carry this out.

The time now is 3:00 p.m. and the bell rings, signaling that it is time to go home. Stephen, Dwayne, and Jerome skip the usual meeting on the street corner to pitch pennies with members of the Cobra gangbangers. They decide to go straight home since Stephen is feeling pretty nervous about his big day tomorrow evening.

While they are walking home, a car pulls up and one of the Cobra gangbangers gets out of the car and walks up to Stephen and tells him they will be by his house tomorrow evening to pick him up around 6:30 p.m., a half hour before the game ends.

He further tells Stephen that they will drive-by the school and once they spot Jason, then Stephen can bust a cap in Jason.

The Cobra gangbanger gets back into the car and the car speeds off peeling rubber. As they continue to walk home, Jerome and Dwayne notice that again

Stephen is very quiet and they ask, "Stephen what are you thinking about man?"

Stephen says, "Did you notice the friendly way in which Jason approached me after recess was over asking me if I was going to the game?"

Dwayne says, "Yeah, he acted as if nothing happened between you two."

Jerome says, "When you bust a cap in him tomorrow, maybe this will jar his memory."

Stephen says, "But that's the whole point, nothing did happen between us."

Dwayne says, "Yes something did, he called you a punk in front of the Cobra gangbangers, the boys we have been hanging out with and want to become some day."

Stephen says, "Yeah, but I called him a punk first, so he was just saying it out of spite. I'm sure that he did not mean anything by it."

Jerome says, "So what are you saying? That Jason did not disrespect you in front of the boys?" Stephen says, "I disrespected him first."

Jerome says, "Being part of the Cobra gangbangers you can do that because we are suppose to be tough. It's expected of the Cobra gangbangers and anyone hanging around them must show that they are tough as well."

That's what being in a gang is all about. I think you need to get your priorities straight, Stephen."

Stephen says, "I have my priorities straight. It's just that I can remember when Jason and I were in the second grade together, I used to go over to Jason's house and his mother would serve us ice cream and other treats and Jason, his sister, and I would play together.

Jason would always have that same smile on his face, just like the one he had on his face when he asked me whether I was going to the game.

It bought back memories of when we were such best friends."

Dwayne says, "Hey, didn't you have a crush on Jason's sister?"

Stephen says, "Yeah, she was such a nice person. Who would ever want to shoot and kill her? That's why when I become a Cobra gangbanger, I am going to find out who shot her and avenge her death."

Jerome says, "Yeah, when I found out that sweet Samantha got shot, I was shocked, surprised, and angry."

Dwayne says, "Rumor has it that a stray bullet hit her from a rival gangbanger's gun."

Stephen says, "I bet that bullet came from one of the Crooks gangbanger's gun."

They now arrive at Stephen's house and Jerome and Dwayne tell Stephen they will meet up with him tomorrow and head out to school together.

Stephen goes into his uncle's house to do some more thinking before his big day tomorrow evening. Jerome and Dwayne head home.

Suddenly a car pulls up in front of Stephen's uncle's home and Little Dan jumps out of the car, runs up to Stephen's place of residence, and starts banging on Stephen's door and ringing the door bell shouting "Stephen, open the door man."

Stephen comes to the door, opens it, and replies, "What's up Little Dan. What's going on dude?"

Dan says, "I need those five bullets that Tyrone gave to you." Stephen says, "What for man? I won't be able to shoot Jason tomorrow with no bullets in my gun."

Dan says, "Tyrone's cousin has just been shot by a rival gangbanger and we have no bullets left."

Stephen says, "Dude, can't you go and see one of your other suppliers that can supply you with some bullets?"

Maybe they can get some bullets illegally from the gun shop and sell it to you guys."

Dan says, "There is no way this is going to happen under the new computerized gun-regulating law called Tracking the Bullet.

Everything is tracked and accounted for since the new gun-regulation law went into effect.

There have already been several suppliers banned from purchasing bullets because they were exposed using bullets they purchased irresponsibly.

As a result of this, we are unable to get anymore bullets from those suppliers due to this new gun regulation law.

Tyrone told me to come and get those bullets that he gave to you since this is all that we have left.

Stephen says, "How am I going to pass my initiation if I cannot carry out the shooting?"

Dan says, "Take it up with Tyrone tomorrow, now give me those bullets."

Stephen goes into the house to get the bullets and gives it to Little Dan. Dan jumps back into the car and takes off, peeling rubber.

Stephen is now thinking to himself how will he be able to carry out the shooting of Jason with no bullets for the gun that Tyrone gave him.

Suddenly he thinks to himself about his uncle's gun. He recalls that the last time that he used his uncle's gun for target practice, he left the gun with bullets in it when he put it back into his uncle's closet.

The next day Jerome and Dwayne meet up with Stephen as they head off to school. Stephen immediately relates what happened the previous day regarding the taking back of the bullets by the Cobras.

He then tells them that he will be using his uncle's gun to carry out the shooting of young basketball player Jason.

As they arrive at school, they notice that they don't see any of the Cobra gangbangers hanging on the street corner pitching pennies as usual.

After school is over, the three boys head home. They notice that they still do not see any of the Cobra gangbangers hanging on the corner pitching pennies.

They could not help but think to themselves whether any members of the Cobra gangbangers got shot while retaliating on rival gangbangers that shot Tyrone's cousin and maybe this is why they did not see any of them hanging around on the street corner.

Suddenly Tyrone pulls up in a car and signals for Stephen to come over to the car. Stephen goes over to the car as Dwayne and Jerome look on.

Tyrone says, "What's up dude?" Stephen says, "What's up? "Did you guys cap those guys that shot your cousin?"

Tyrone says, "Pretty much." Stephen says, "Listen, I can get my uncle's gun to carry out busting a cap in that punk Jason to complete my initiation."

Tyrone says, "We will come by to pick you up around 6:30 p.m, bring your uncle's gun."

Tyrone drives off, peeling rubber, as Stephen rejoins Dwayne and Jerome, as they continue walking home. The three have now returned to their place of residence.

Several hours pass by, and the time is now 6:30 p.m., and Tyrone pulls up in the car to pick up Stephen. He honks for Stephen to come out.

Stephen grabs his uncle's gun out of the closet and goes outside and gets into the car with Tyrone; as he gets into the car, he notices that there are two other Cobra gang members in the car with ski masks in their hands to hide their identity.

The car peels rubber as they head out to go and commit the gun homicide against young basketball star Jason.

The time is now 7:00 p.m., and the basketball game between the Carter wildcats and the Powell warriors has ended. Students, friends, and faculty head outside from the gym room auditorium to go home.

Tyrone pulls up in a black Chevrolet Sedan with tented windows rolled up as they search for Jason to come outside. They see members of both basketball teams coming outside from the gym room auditorium of Carter elementary school.

Coincidently, detectives Sharp and Jones are within the area, they are responding to an earlier call that they received over the car radio regarding shots fired near Carter elementary school.

They are cruising the neighborhood in an unmarked squad car searching for the perpetrators. They are in a 2005 dark blue Chevrolet Impala.

Finally one of the Cobra gangbangers who is inside the car spots Jason walking down the street with Sabrina, heading home. He then says, "There's Jason." Tyrone asks, "Where?" Stephen says, "Straight ahead, walking past that yellow two-story building." Tyrone tells Stephen to get ready.

Detectives Sharp and Jones turn the corner, they are now on the same street as Tyrone and the gang members inside the car, just two-car lengths behind them. They notice that the car looks suspicious and detective Sharp radios in to get information on the license plate number.

Meanwhile, Tyrone is unaware that the detective car is behind them; he then speeds up to catch up with Jason. Detectives Sharp and Jones trail them. Tyrone pulls up to Jason and Sabrina and the back tinted window rolls down slowly.

Stephen, wearing a ski mask, sticks his head and right arm out of the back window, aiming directly at Jason. Jason and Sabrina see the young man wearing a ski mask stick out of the window pointing a gun at them. Jason tells Sabrina to run in the opposite direction. Sabrina immediately starts running down the street in the opposite direction screaming for help.

Detectives Sharp and Jones notice this. Jones replies, "Hey, that's my daughter." Meanwhile, Jason runs two houses down in the opposite direction from Sabrina and ducks in a residential gangway and starts running through the gangway, trying to escape being shot. He knows that the car cannot jump the huge sidewalk and fit into the residential gangway.

Stephen immediately gets out of the car and starts chasing after Jason toward the gangway. Detective Jones jumps out of the squad car and tells his daughter to go back into Carter Elementary School for safety purposes. He then starts to chase Stephen, yelling, "Stop and drop the gun, this is the police."

Meanwhile Tyrone notices that the police are now involved and he speeds up to try and get away. Detective Sharp is right on his tail.

Detective Sharp radioed in for backup, while detective Jones continues to chase after Stephen through the residential gangway, yelling, "Stop and drop the gun."

Stephen ignores the warning and continues to chase after Jason through the residential gangway.

Suddenly Jason trips and falls down on the gangway ground. Stephen catches up with him and stops and stands over him.

Jason looks up at the guy wearing the ski mask and tells him, "Please man, don't shoot, don't shoot." Stephen aims the gun and pulls the trigger, but no bullets come out. He pulls the trigger several more times and still no bullets come out. Every time he pulls the trigger, Jason flinches for fear of being shot.

Detective Jones is now only a few feet behind Stephen and he yells out loud, "Freeze! Drop the gun now and put your hands up into the air."

Stephen realizes that the gun is empty and he is now in a no-win situation; so he complies with detective Jones demands and drops the gun and puts his hands up into the air, as instructed.

Detective Jones comes up behind him and puts the handcuffs on Stephen. He then asks Jason, "Are you all right young man?" Jason replies, "Yes officer."

Detective Jones helps Jason up off the ground and then he turns to facing Stephen and pulls the ski mask off Stephen's face. Jason looks and notices that it's Stephen, his second-grade school pal.

He says to Stephen "Hey man what's up, why were you trying to shoot me?" Stephen just hangs his head down, feeling ashamed. Detective Jones replies, "You two know each other?"

Jason says, "Yes, he and I were best friends in the second grade, but I don't understand why he is trying to kill me." Detective Jones looks at Stephen and says, "What about it young man, why were you trying to kill him?"

"I saw and heard you pull the trigger several times, so don't even try to deny it." Stephen starts to cry and said that he was trying to pass the initiation to make it into the Cobra gang.

Detective Jones, Stephen, and Jason walk back to meet Detective Sharp at the squad car. Jones says to Jason "I'll need you to take a ride down to the police station to press charges. We will contact your parents when we get down there."

They now arrive at the police station. While at the police station, sitting down, Stephen notices that Tyrone and the two other Cobra gangbangers that was in the car with him are being taken into custody with handcuffs on their wrists.

Detective Jones contacts Jason's parents, Denise and David Williams. He also contacts Stephen's guardian, his uncle Mike, to come down to the police station.

Mike arrives at the police station first. Detectives Jones and Sharp take Mike and his nephew into a side room and began to question both of them. Jones questions Stephen as to why this happened and where did he get the gun and whose gun is it and how many times has he used the gun and if he could remember the dates that he used the gun.

Stephen answers all of his questions. Jones then questions Mike and wants to know why his gun was empty this time as opposed to the other two times when Stephen and he shot the gun.

Mike explains that he tried to buy more bullets for his gun, but due to the new gun regulation law he was denied for failing to account for one of the six bullets that he previously purchased.

Once Denise and David arrive, detectives Jones and Sharp take them into one of the side rooms and tell them everything that happened and relate to them Stephen's confession on trying to shoot their son, Jason.

Denise replies, "You mean to tell me that this other boy named Stephen would have shot my boy to death had there been bullets loaded into his uncle's gun?"

Jones says, "Yes, but because his uncle was unable to purchase more bullets for his gun due to irresponsible usage and accountability under the Tracking the Bullet gun regulation law, your son is alive today."

We also believe that the reason why your daughter was shot was because of a stray bullet fired from Stephen's uncle's gun, either by Stephen or his uncle.

They live two blocks away from you, opposite area, but facing the same direction from your back porch where Samantha was hit by the stray bullet.

Stephen confessed to the first time that he shot the gun was in the alley around the same time that your daughter was hit, while showing off in front of his two friends by firing several shots up into the air.

One of those stray bullets could have hit your daughter while she was standing up, not sitting down, on the back porch.

This would explain why she fell over the back porch banister into the backyard grass. Had she been sitting down when hit, she would have fallen down on the back porch floor, up against the porch banisters.

Also, his uncle two days later, on the Fourth of July shot several shots up into the air. All of this happened within the same time frame of Samantha's death.

Although your daughter's death has now been changed from a homicide to an accidental incident, we can still press charges for illegal use of a firearm. It is illegal to fire shots randomly up into the air.

Before the new gun regulation law, there was no way to actually prevent a person from being a repeat offender or having the same gun-related incident happen over and over again until now, but thanks to the new gun regulating law, this is possible."

Denise says, "Tracking the Bullet saved my child's life, and I am very grateful. Thank you so much. May we see our son?" Sharp says, "Yes, he's just across the hall in the other room. Come on, I will take you to him."

As they walk across the hall to meet Jason, tears of joy are falling from the parents' eyes. Denise and David give Jason a big hug, as they are happy that he is alive, thanks to the new gun regulation law called Tracking the Bullet to Save Lives.

Detective Sharp goes into the side room to let Mike know that his gun has been confiscated pending illegal usage charges and that his nephew Stephen has attempted murder charged against him and that he may be tried as an adult depending on how the district attorney wants to handle this.

Mike now realizes why he only had five bullets left in his gun at the firing range instead of the six that he previously purchased.

He is also very thankful for the new gun-regulating law that prevented him from purchasing more bullets for his gun due to his neglect of not being responsible for what happened to all six of his deadly bullets.

He feels bad for his failure to keep his gun out of the hands of a minor. He realizes that if he would have purchased more bullets for his gun, Jason might not be alive today. He is also thankful for the new gun-regulating law called Tracking the Bullet to Save Lives.

He also wished he would have suspected and questioned his nephew about this earlier, but he never suspected his nephew could do a thing like this.

He feels that he could have prevented all of this from happening had he paid more attention to his nephew's activities and his whereabouts.

Chapter 3

A True Story

The story that you just read had a very happy ending. A young student's life was spared due to tracking deadly bullets that kill. And being a young basketball star, he can now grow up to be the next Michael Jordan, a basketball star.

However, the truth of the matter is that stories similar to this happen all the time in this country, where innocent lives are lost due to irresponsible usage of a firearm. Many students do not get the chance to grow up to be that next basketball star or that next scientist, engineer, fireman, or something else great. Their stories do not have a happy ending due to gun-control laws that are not setup to address the real problem at hand today.

Let's stop and reflect for a moment about the little girl name Samantha who was hit by a stray bullet that landed in her heart, while standing on her back porch enjoying the summer breezy day.

This could have been a new born baby being carried in her mother's arms and while coming from the hospital, down came a stray bullet that launched right into the temple of that child's head and instantly killed the infant. The very fact that incidents similar to this keep happening over and over again needs addressing to say the very least.

We don't take the time to reflect on what stray bullets can do to a person because it does not make the 9:00 p.m. news. We do, however, reflect on the drive-by shootings by gangbangers, which are even worse because there is a targeted aim. What does all of this tell you?

That it's time for a change. We need to take back control of the way a person uses his gun. If not, that daughter, that son, or newborn baby could very well be yours—the next victim that a deadly bullet claims.

The latest data from the United States Centers for Disease Control and Prevention show that 3,184 children and teens died from gunfire in the United States in 2006—a 6 percent increase from 2005. This means one young life is lost every two hours and forty-five minutes, almost nine every day, sixty-one every week.

Of these deaths, 2,225 were homicides, 763 were suicides, and 196 were due to an accident or undetermined circumstances. Boys accounted for 2,815 of the deaths and girls for 369 deaths. More than five times as many children and teens ̄17,451 ̄suffered nonfatal gun injuries.

- The number of children and teens in America killed by guns in 2006 would fill more than 127 public school classrooms of twenty-five students each.
- More preschoolers (sixty-three) were killed by firearms than law enforcement officers (forty-eight) killed in the line of duty.
- Black males aged fifteen to nineteen are almost five times as likely as their white peers and more than twice as likely as their Hispanic peers to be killed by firearms.
- Between 1979 and 2006, the yearly number of firearm deaths of white children and teens decreased by about 40 percent, but deaths of black children and teens increased by 55 percent.
- Since 1979, gun violence has ended the lives of 107,603 children and teens in America.
- 60 percent of them were white; 37 percent were black.
- The number of black children and teens killed by gunfire since 1979 (39,957) is more than ten times the number of black citizens of all ages lynched throughout American history (3,437).

The United States remains one of the few industrialized countries that place so few restrictions on gun sales. There are more than 270 million privately owned firearms in our country—the equivalent of nine firearms for every ten men, women, and children.

The daily news is a grim reminder of the devastating impact caused by our deadly romance with guns and violence. What will it take for us to stop this senseless loss of young lives? Individuals and communities must act to end the culture of violence that desensitizes us to the value of life.

End quote from the United States Centers for Disease Control and Prevention.

With these statistics alone, we can see that there is a need for change. Because even with the toughest gun-control laws in place today, you still have gun homicides rising at an alarming rate.

We need gun-control laws that can control how a person uses his gun, not just for the consequences that he faces if he's caught using it illegally. By then, the harm has already been done.

The gun-control laws in place today are geared more toward restriction and not prevention. The current control laws control how much time a person will spend in jail if he is caught illegally using that gun.

It's time we stop being so passive about our kids' future and start taking back control. We can start by holding a person accountable for every bullet that he purchases. Once we begin to track how he is using those bullets, we then start to control how he uses his gun.

In the story you just read, bullets cost only a few dollars; it took only one of those bullets to claim a life whose value is worth much more.

Can the owner of the gun account for where, when, and how those bullets were used? These are questions that we should be asking responsible adults.

A responsible adult is the person who has the legal right and the qualifications to purchase bullets for his or her gun. A minor does not have those qualifications or rights.

So when a responsible adult purchases bullets for their gun, they now become the owner of dangerous pieces of metal. As owners, they now become accountable for how those bullets will be used.

Accountability can only be enforced through tracking how the gun is used. When you track the bullet usage, then you can start to control the gun.

Once you start to control the gun by tracking a person's bullet activities, you would then start to save lives by eliminating irresponsible, reckless usage.

Without controlling the bullet, you are not controlling the gun because the whole purpose for the gun is to shoot those deadly pieces of metal called bullets. Once you lose that control, you now have an uncontrolled situation and the results can be deadly.

Take for example the owner of a pit bull dog who took it out for a walk and someone was walking on the opposite side of the street and the dog broke loose from the owner's control and attacked the person on the opposite side of the street and mauled that person to death. Who would be responsible for the person that is now dead from being mauled to death by the owner's dog: the dog or the owner of the dog? The answer is obvious. The dog was under the owner's control or should have been. He broke loose from the owner's control. The owner did not exercise his control properly. Therefore, it makes the owner of the dog liable.

The next important question would be should the owner of that pit bull dog get another chance to repeat this fatality, which could claim someone else's life due to the negligence in failing to control the dog? However, if he is not held accountable for what happened by revoking his privileges to continue to own that vicious dog, you will have the same situation where this process can be repeated over and over again. Other people may lose their lives because of this owner's negligence in failing to control his vicious dog.

That's what is happening with the out of control gun usage that is claiming innocent lives today. You have repeat offenders out there who never get caught or stopped in their tracks because there are no gun-control laws to address the situation that we are faced with today.

For something so deadly and dangerous as a bullet that can claim a life, shouldn't there be accountability on just where, when, and how these dangerous pieces of metal are being used? I mean, think about it, we do this for hazardous material. Then why can't we do this for bullets being fired out of someone's gun, which is just as dangerous, if not more, as the hazardous material, depending on the circumstances.

If you ask any freight train conductor who is in charge of the transportation of the freight on a freight train, he will tell you that there is a tracking trail done for each hazmat traveling on a freight train. There is paperwork that explains where this hazardous material is going and when and how this hazardous material will be transported.

It even tells you what type of hazardous material it is and the emergency phone number to call in case there is any spill.

So this concept of tracking the bullet through bullet control mechanism that is being proposed here is already being used on other equally dangerous commodities such as hazmats.

Why not use it on bullets as well by making the owner accountable for their bullet usage. The system that I am proposing will track a person's bullet usage.

In time, you will have the answers to the question, how did those bullets end up in the hands of a young gangbanger?

Let's prevent this from happening through a gun-regulation system that can virtually stop gun violence by tracking the bullet.

In most cases you will find that the neglect first started with the person who has the legal right to purchase those bullets.

How those bullets transferred from the purchaser's hands and ended up in the hands of that young gangbangers gun, who then shot and killed someone, can be determined and prevented.

This is what the Tracking the Bullet concept is all about. Prevention is the key. When you can prevent a potentially dangerous situation from escalating out of control, you have regained control of an otherwise out of control situation.

The problem first started with the purchaser of the bullet not being held accountable as to what he or she is doing with those deadly bullets that are purchased.

We have to address this just like we address the dangerous hazmat commodity for safety to the public.

Supplying these bullets to these young gangbangers is an act of irresponsibly using bullets that kill. When we take this irresponsible usage out of the equation, we start to save lives in the process.

With the computer technology that is out here on the market today, we can use this for the betterment of society. With this technology, we can implement a system that can actually track a person's bullet usage.

Doing so will create more jobs and this in turn will boost the economy as well. So let's recap for a moment here. Save lives, create more jobs, and stimulate the economy as well. Now that's a win-win-win situation if I ever heard of one.

The gun control laws that are in place today are not getting the job done. If they were, you would not have the statistics constantly going up on gun-related crimes.

If a person is using his bullets responsibly and he can prove this, then nothing will change for that individual other than the fact that he will have to prove this.

It is the irresponsible usage of one's bullets that I am proposing that we fix. The statistics of gun-related homicides will then go down. And really, wasn't that supposed to be one of the original purposes for the gun-control laws that have failed so miserably in Chicago alone.

What good is a law if it does not serve the intended purpose for which it was enacted? You first have to look at the root cause before you can begin to try and solve the problem. One of the biggest problems is a moral one.

In this country, we have become so concerned about making money that we are indirectly diminishing the value of life. I mean, let's face it, guns are big business; but then again, so are drugs, alcohol, and tobacco.

And these can be some of the root causes of some of the most hideous crimes that happen today. Tracking the bullet is taking a step in the right direction.

I'll be the first to say that as parents it is our responsibility to bring up our children in a manner that they learn the qualities of discipline, respect, and love so that they can grow up to become responsible adults and contributing citizens, contributing positively to society and not negatively. The reality though is that all too often this does not happen. The parents of the children that are out there gangbanging often times have not had the proper upbringing themselves and so the domino effect happens.

If the parents have not had such values as respect, love, and discipline instilled in them, how in turn could they pass these values down to their children? And so the values of life have diminished. And through each passing generation, it starts to decline until it hits rock bottom.

The result is that you now have preteens and teens running around with a gun in their hands, shooting bullets that take lives.

If the parents could not raise the child, then the streets surely would. And the streets are full of guns, drugs, and violence.

The youth of Chicago are dying at an alarming rate. Not because they enlisted into the armed forces where they took an oath to serve and protect their country from terrorists. They are losing their lives in the streets of Chicago through gun violence perpetrated by other youths.

No doubt everyone is familiar with the recent headlines of gun-related deaths in Chicago alone. Perhaps some of the young ones that lost their lives are your sons and daughters.

If so, you have my deepest condolences. I hate to think of the mental pain that I would suffer had it been my child who was gunned down in the streets of Chicago.

You work so hard to invest so much in your child, and to have it all taken away at the pull of a trigger and the release of a bullet in the hands of an irresponsible person is not only devastating, but it also makes no sense when you think about how and why this happens over and over again.

And these are only the stories that we here about on the news. What about the gun-related homicides that are happening that don't even make it to the news and are also stories of young ones losing their lives.

One may argue that this increase in gun homicides is because there are too many guns exported into this country. However, this is certainly not an argument to use when it comes to a gun-control law that is supposed to prevent a gun homicide from happening. I mean, think about it, it only takes one gun and that same gun can last for years before it breaks down.

This means that you can keep on reusing that same gun over and over again to commit gun homicides without exporting anymore guns into this country.

When are we going to stand up and start to take responsibility for our role as concerned citizens in this country? Many of us want to, but simply put, we don't know how to.

This book will show you how to do so. The policemen, the detectives, the FBI, and all the law enforcement agencies, all have their hands full in fighting crime.

Can't we make their lives a little easier by narrowing down the culprits who are irresponsibly using bullets that kill?

Although the person who purchased the bullets may not be the actual person who pulled the trigger, he acted irresponsibly by not controlling the bullets that he purchased allegedly for his gun and allowed those bullets that he purchased to end up in the hands of an adolescent.

And often times this is a teenager out there gangbanging and taking lives. We as a society have become so keen on prosecuting the person who actually pulled the trigger and murdered someone, that is, if he's caught.

We overlook the other culprit. If it's a minor that did the killing, such as a teenager or an eleven year old, we should be asking where did he get the bullets for the gun? We already know that there are too many guns out here and that it would be very easy for an individual to get his hands on a gun.

The ammunition, however, is a different story. My system will tract this and at the very least, narrow it down so that law enforcement agencies can focus their investigation at a more targetable location of where the bullets are coming in from.

If the bullets are being exported in illegally, they can focus on how this is being done because the gun-regulation system will virtually eliminate local illegal supplying of bullets to thugs and gangbangers within the United States.

It will do this by making the person who is the source, the one purchasing the bullets and supplying these gang members with the ammunition, accountable for his actions, just like in the story that you just read about in Chapter 2.

This is what I am proposing in this publication. I suggest that we put in place certain mechanisms that would make a person accountable for where, when, and how his bullets are being used. With the technology out here on the market, this can be done.

The second amendment of the constitution states that a person has the right to possess arms to protect himself. That should hold true for only responsible individuals who are using their pistols within the confinements of the law.

While my proposal is not a cure all to gun violence, it can at the very least save innocent lives and solve some of the gun-related homicides that are happening all over this country, especially of late here in Chicago, where gun-related deaths have grown in the last two years alone at an alarming rate.

It would also make law enforcement agencies' life a little easier in tracking gun-related activities once the mechanisms are put into place and have been perfected through trial and error.

Anything worthwhile that is going to bring about a good change takes time, practice, and trial and error. But once the change has been perfected and tested through time and trial and error, it brings forth beneficial results.

The benefits of gun regulation by tracking the bullet is desperately needed and if not put into place soon, the situation is only going to get worse. Look at how these sad headlines read.

CBS Evening News
Shock Over Chicago Student Gun Deaths.

In this school year alone, twenty-four students were killed, leaving the City Rattled to the Core.

Chicago, April 22, 2008:

(CBS) Public school students in Chicago aren't as worried about making the grade as they are about making it home alive, CBS News correspondent Cynthia Bowers reports. "You can't go nowhere without being shot," said Juston Gant "It's Crazy." Since September, twenty-four students have been murdered, most of them shot.

The dead amount to a classroom of kids. Among them was ten-year-old Arthur Jones, who was on his way to buy candy when he got caught in gang cross fire. As did fifteen-year-old Miguel Pedro, who went out for ice cream and never came back.

Last school year, thirty-four students were killed. That's fifty-eight deaths over what amounts to a seventeen-month period. And that makes an average of one child getting murdered every eight days. In a city where handguns are already banned. Stop the violence rallies have had little impact and the mayor's calls for stricter gun laws have fallen on deaf ears.

Angry parents and concerned citizens march the streets, feeling helpless and hurt as they form rallies with cries for help, while this is admirable that they openly express concern for their child and their own safety as well.

They feel powerless and hopeless because they know that something more concrete is needed to stop the gun violence in Chicago and across the nation.

You may be asking yourself, is my marching in the streets having any real impact on these young people taking lives?

In a few days the gun violence will start up again because the young criminals know that they have the upper hand as long as they are being supplied with the ammunition for their guns.

However, out of concern for your child's welfare you speak out about how the government should be passing tougher gun laws to cut down on some of the gun-related homicides like young people being gunned down in the streets of Chicago and other large metropolitan areas.

Tougher gun laws have been passed such as the banning of handguns; yet, you still have gun-related homicides that are happening all over this country.

The problem is that hardened criminals have neither respect or regard for the law nor do they abide by it.

Placing a ban on certain types of guns has little impact on preventing a person from using his gun irresponsibly or illegally, especially since there are so many guns already out there on the streets.

The criminal element is not going to roll over and play dead just because a law has been passed banning handguns in Chicago or other large metropolitan areas. We need to be working on prevention.

I mean, think about it, what gun law can you have the government pass that can really have an impact on gun-related homicides without violating a person's rights to possess arms?

That is, until now. With the Tracking the Bullet gun-regulating system put in place, it certainly would not violate a person's right to possess arms, it only targets how those arms are being used. How? By targeting the source as someone responsible or irresponsible.

Let's take a look at the qualifications needed to purchase firearms and ammunition in the state of Illinois. In order for a person to purchase a gun and ammunition for his gun, he would need a firearms identification card also called (FOID).

So he would need to apply and register for the FOID card. During the registration process, this person has to pass a background check and he cannot be a criminal or a former criminal. So obviously, a person who has a criminal record would not qualify for a FOID card and rightfully so.

Does this stop this person from getting his hands on a gun and some ammunition? With the amount of guns that are already out there on the streets, he has plenty of opportunities for finding himself a gun.

There is always a person out there who will sell him a gun illegally to make a fast buck.

Everyone is always harping on gun violence, tougher gun laws, guns kill, but nobody hardly ever says anything about the bullet; yet, it's the very thing that the coroner pulls out of the body of the slain victim when trying to find out what type of bullet took this individuals life.

A gun is only as good as its bullets.

If you have lots of fancy looking guns out on the streets with no bullets in them, what good are they if you can't fire a shot.

So it just makes sense to track the bullet usage, which ultimately causes the death, rather than focusing so much on the gun.

So how bad do you want to save your child's life from gun violence?

Obviously you want to really badly; otherwise why are you marching in the streets of Chicago and other metropolitan areas forming rallies and pushing for tougher gun laws.

I have revealed to you with convincing evidence why the current gun-control laws that are in place do very little to stop gun violence. The criminals do not respect those laws and continue their efforts to seek and destroy through gun violence by shooting and killing innocent people.

There is nothing preventing this new breed of young murderers from killing innocent people. They are young adolescents killing other young adolescents in the streets of Chicago and throughout the nation.

These new breeds of murderers today do not know or care about the value of life.

They are fearless and careless in their conduct. They do not think about the victim that they are targeting to murder or the victim's loved ones or how the loved ones will feel once they learn that the person who was shot and killed was a relative.

When we take a look at the gun system and how it works today, we can then start to categorize what needs fixing and what doesn't.

There are people out there who use their guns for sports such as hunting or go to the firing range for target practice. These people are using their bullets responsibly without endangering human life.

These people are not the ones that the gun-regulation system will fix. They are already regulated.

The law enforcement agencies use their guns to serve and to protect law-abiding citizens through a sworn oath. They also are not what the gun-regulation system will fix.

They are already held to a higher standard. It's the irresponsible ones out there that we need to regulate and fix.

We can't keep assuming that most of the gun-related homicides are the results of bullets coming into this country illegally because we see movies depicting this and this is how the gangbangers get their hands on bullets that kill.

They all can't be receiving them from sources overseas illegally into this country. And yet your local legal gun shops remain in business, doing quite well.

The source is a local supplier. He or she is an irresponsible individual who has the legal credentials and qualifications to purchase bullets in this country and then turn around and illegally provide these bullets to the criminal elements within the United States.

This local supplier is undoubtedly financially well off due to the supply and demand scenario, but then again, so would the local gun shops that he or she is purchasing bullets from.

Since the gangbangers do not have the legal credentials and qualifications to purchase bullets legally themselves, they rely heavily on this local supplying source to provide them the necessary ammunition for their guns.

And given the number of lives that these young gangbangers are taking with these deadly bullets every day, the supply and demand for more bullets is big business.

Think about how many people in this country lose their lives every day from gun violence? Have we become so naive that we think that the bulk of this activity is due to bullets coming into this country illegally from overseas and then being supplied to the criminal element?

If that were the case, then the gun shop industry would have gone out of business a long time ago in this country. They would no longer be needed since the bullets are being supplied from overseas illegally. And the supply is so abundant that the gun shop owners are unable to make enough sells on their bullets within their gun shops to survive.

So ask yourself this question, how are the gun shops staying in business if the bulk of the business is coming from overseas illegally?

Remember, guns can last a long time and the same gun can be used over and over again, but bullets however, you have to keep on repurchasing once you use them up. It's much easier, cheaper, and convenient to get them locally rather than overseas illegally.

For any business to survive, there has to be supply and demand for the product that the business sells.

Gun shops sell more bullets than guns simply because guns last longer than bullets. Bullets have to constantly be repurchased because you cannot reuse the same bullets constantly like you can use guns.

So there has to be enough bullets sold in order for that gun shop to keep on making a profit so that it can survive. So the supply and demand scenario is definitely there, especially in Chicago alone with skyrocketed gun homicides on the rise.

Currently, it is not the gun shops' responsibility to track what you are doing with your bullets once you have purchased them from the gun shops.

Their only responsibility is to make sure that you have the legal qualifications and credentials to purchase the bullets for your gun and once you do that, it's out of their hands.

You could be one of hundreds of individuals purchasing bullets from the gun shops legally and reselling those bullets to the criminal elements illegally without no accountability whatsoever as to what you are doing with the deadly bullets that you purchased.

The gun-regulation system called Tracking the Bullet to Save Lives will change this.

If we had bullet sensors that we could place on every bullet that is sold and used to keep track of where that bullet eventually ends up by means of a computer software-tracking device located at a central location.

Now that would be really cool. I mean, think about it, the sensor on the bullet would be able to tell you whose gun it has been place in, the owners name, address, and the location of where it has been fired.

Unfortunately, we are not that technologically advanced where we can do this yet, but maybe we could in the future.

We do have technology out there that can be implemented to track a person's bullet usage, which will take some manual processes as well, which is where additional employment will be needed.

The computer technology that we currently have on the market will help out a great deal in making this plan work.

There is only one real showstopper that is standing in the way right now, and that's where you come in—making this gun-regulation system work.

In the story that you read in Chapter 2 of this book, you saw how Denise took the steps to make this a state and federal law. Follow this example when pushing to make Tracking the Bullet to Save Lives a gun-regulating law, granted it may take much longer than three weeks to have it passed as a state then a federal law.

When it passes, you will start to see gun violence decline considerably. You will start to take back control of uncontrolled gun usage that is claiming innocent lives at the hands of irresponsible ones.

The gun-regulation system called Tracking the Bullet will save lives, it will create jobs, and it will stimulate the economy as well.

You now have within your power to help push to make this a state law first and then push to make this a federal law. Either way, it will cut down tremendously on the gun violence here in Chicago and other large metropolitan areas when made a mandate.

Chapter 4

How the Gun Regulation System Works

How do we track the bullet? Simply put, by making a person accountable for what he or she does with the bullets that they are purchasing through a recertifying process. Every time they buy bullets for their guns, they are in fact certifying that they have used their bullets responsibly the last time they purchased bullets for their gun.

Once they pass the certification process, only then will they be granted the privilege of repurchasing bullets. If they fail the recertifying process, they are taken out of the bullet purchasing system forever because obviously they have shown that they have used their bullets irresponsibly.

Recertifying is a broad usage term. When you think of the term what comes to mind? We recertify every time we go and renew our driver's license. We are in effect saying that we want to continue to have the privilege of driving on the streets among our peers in a safe manner and we will do everything possible to prevent causing an accident by driving safely.

And so we are asking the state to renew our license after we have proven that we are qualified to continue driving responsibly.

We are willing to prove that we are fit to continue driving our automobile. We are given a rules of the road test and an eye test to check the competency of our claim that we are fit to continue driving responsibly.

Our answers to the rules of the road test are then verified against a master score sheet or database to see if we entered the correct answers, verifying if we should or should not be granted the privilege to continue driving based on our answers.

If we fail the eye test, then we would have to go and get eyeglasses and retake the test again. If we fail the rules of the road written test, then we would need to study more until we get it right.

Recertifying to purchase bullets for your gun is similar in nature. If you fail to account for all of the bullets that you previously purchased, then you are denied the privilege to purchase bullets for your gun until you can responsibly account for those bullets that you purchased.

And just like the rules of the road test, the answers that you provide on the recertifying application will be checked against the database to see if you have used your bullets responsibly or not.

The only difference here is that if you fail to account for those bullets by virtue of showing that you used them irresponsibly, then you are permanently denied the right to purchase bullets again.

The system locks you out so that you will not be given the opportunity to continue being an irresponsible user. Your irresponsibility can cost someone else his or her life.

A person must first pass a background check to see if he or she qualifies for a firearms identification card or FOID. He or she is then sent an electronic swipe card along with his or her FOID card. If a person lives in a state where a FOID card is not a requirement to purchase a gun, they will still need a swipe card with a personal identification number for the purpose of being regulated under the Tracking the Bullet gun law.

If he cannot pass the background check, then he cannot even receive a FOID card. For those who already have a FOID card, they will be mailed a personal swipe card with their personal identification number stored into the magnetic strip on the back of the swipe card. They must recertify every time they want to purchase bullets for their gun.

This swipe card is similar in nature to a bank debit card. It has a magnetic strip on the back of the card for swiping through an electronic card reader. The card reader is attached to an electronic computer cash register. The electronic computer cash register is connected to a high-speed communication line such as a DSL, or a T1, or a fiber optic high-speed communication line.

The computer cash register connects online through the high-speed communication line to a centrally located computer database server.

The purpose of this computer database server is to send and receive data to and from the software database located on the computer database server.

For the purpose of tracking your personal bullet usage, the card reader interfaces with the software database through the computer cash register.

The card reader would read the magnetic strip and send that information to the software database for tracking your bullet activities. The software database would then send data back to the computer cash register asking for your personal identification number.

There will also be an electronic keypad attached to the computer cash register for entering your personal identification number or PIN. The electronic keypad interfaces with the software database through the computer cash register.

When a customer enters his or her personal identification number using the electronic keypad, that information is transmitted back to the software database.

The software database would then received this information and check this information against the information stored in the software database for this customer.

Once the information is verified, the software database will come back with a series of questions for this customer that will need answering in order to determine if this customer should be granted the privilege to repurchase bullets again or not.

The clerk would enter the answers to these questions using the computer cash register keyboard. The clerk will get the answers to these questions from the certifying application that the customer will fill out when he or she is trying to repurchase bullets.

Recertifying involves filling out an application and swiping your swipe card through the electronic card reader attached to the computerized cash register.

Here's how it would work. The first time you purchase bullets for your gun, nothing will change except that the system will track your bullet purchases when you use your swipe card to keep a tracking record of what you did with those bullets the next time you come in to purchase bullets again.

The swipe card and pin number is similar in nature to the bank debit credit card and pin number that you would receive from your bank for your checking account.

When you go to the bank and swipe your card, you are required to enter your personal identification number or (PIN) proving that you are the rightful owner to access this bank account that this bank debit card is drawn off.

The swipe machine would be similar in nature to what is used at the cash register in a grocery store. The gun shop store clerk will be able to validate if you can repurchase bullets for your gun or not.

Let's say that you have used your bullets responsibly at the firing range or at a deer hunting site and the computer was able to track every bullet that you purchased and used at the firing range or deer hunting site. The repurchasing process would be painless and simple.

You would simply go up to the cash register and swipe your card to purchase more bullets. The system will have in the database a tracking record that you have used every bullet that you last purchased responsibly thereby letting you purchase more bullets for your gun immediately.

When you first use your swipe card to purchase bullets for your gun, the gun shop store clerk will be able to validate based on a two-digit code that the computerized cash register returns from the centralized computer database.

The technology here is similar in nature to when you go and purchase something with your debit card. When you swipe your debit card into the machine, you enter your personal identification number, and that data is sent to a database of the bank holder of your debit card account number. If you have the funds available in your account to make the purchase, then the data comes back approved and you can purchase the merchandise.

You would swipe your FOID swipe card through the computerized swipe card machine attached to the computerized cash register at the gun shop where you purchase bullets.

You would then enter your personal identification number for your FOID swipe card. The computerized swipe card machine would then send the data back to the centralized computer database that is located on the centralized computer at the local police station.

That computer at the local police station is just one of hundreds of computers throughout the country that share the same database information. It will check the database to see if you have purchased bullets before. It will return back with a code of 01, 02, or 03. If it returns a 01 code, this means you did not purchase bullets before, so you are allowed to purchase bullets for the very first time that you used your swipe card with no problem.

If it returns with a 02 code, this means that you have purchased bullets before and now you would have to recertify in order to purchase bullets again.

Along with that 02 code, there will be an explanation showing if you used the previous bullets that you purchased responsibly or not by having or not having a tracking record of all your last purchased bullets.

Again if you used your bullets at the firing range or a hunting site, then the system would have tracked every bullet and you are granted immediate access to purchasing bullets again since all bullets have been accounted for. You have shown that you are a responsible customer.

If the explanation comes back negative, meaning that every bullet that you last purchased has not been tracked and accounted for responsibly, the store clerk would see this and will let you know that in order to purchase bullets again, you will need to recertify showing that you used those last bullets that you purchased responsibly.

The clerk will then hand you a recertify application to fill out. The answers that you provide to the questions on the certifying application along with your supporting documentation proving that you used your gun responsibly will be verified by the tracking computer database application software to see if you have used your previous bullets that you purchased in a responsible manner. If you have, once the verification process has completed, which should take no more than a couple of weeks, you will then be sent a letter in the mail from your local police station letting you know that you have passed the verification process on your certifying application and now you can repurchase bullets for your gun.

When you go back to the gun shop you will go through the FOID card swipe process again to purchase bullets for your gun. The electronic system will then return a code of 03, which means that you have passed the recertifying process and may repurchase bullets again.

Once the repurchasing process is completed and you have purchased bullets for your gun, this will be tracked by the database.

The system will send this information back to the centralized computer database, overwriting the 03 code with a 02 code for the next time when you want to purchase bullets again.

So when you have used up all of those bullets that you've just purchased you return back to any gun shop to purchase bullets again, you will need to go through the process of certifying again, proving that you used your bullets responsibly since the last purchase you made.

The questions on the certifying application are tailored in such a way that one can see if you have used your bullets responsibly. The most important questions are. 1. "How have you used the bullets that you last purchased?" 2. Where have you used the bullets that you last purchased? 3. When did you use the bullets that you last purchased and how many bullets did you use? The answers that you give to these questions will be verified to show if you have used your bullets responsibly or not.

So the correct answers that you can possibly give to show that you have used your bullets responsibly or not will be 1. "You used the bullets for target practice at the firing range or at a deer hunting site." 2. "You went to the firing range or deer hunting site at this address." 3. "You used the bullets on this date and time and the total number of bullets used was." If all goes well, the computer will have a tracking record of this. If one bullet goes unaccounted for, you are denied access to purchasing bullets until you can responsibly account for that one bullet.

Keep in mind that the answers to these questions will be verified by the computerized database.

The firing range will also have a computerized cash register with an electronic card swipe machine attached that interacts with the centralized computer database.

The clerk will also be responsible for checking your gun for the number of bullets in your gun when you arrived and when you leave. All of this information gets entered into the computer by the clerk. They will also have metal detectors within their facility that you will have to pass through when you arrive and when you leave.

This is for the purpose of tracking any bullets that you may have on your person. If they find any, they will enter this number in the system as well.

The metal detector machine will be similar to the metal detector machines that are located and used in the federal building courthouses.

If you give any response other than the correct answers that the database system is looking for or your answers to the questions can't be verified by the database system, then you will have to send in supporting documentation backing up your story as to what you did with the bullets proving that you used your bullets responsibly. And this too will be checked for verification of truth.

As an example, if your answer to question number one was that you used your bullets defending yourself, then you will have to explain this in the added space on the application.

Let's say that someone broke into your home and tried to rob you and your family and you used your gun to defend yourself, shooting this person and you recorded this on the certifying application and handed it to the clerk.

You will be sent a letter asking for documentation to verify your story such as a police report that this actually did happen.

Now let's say that you send in your police report with the verifying application and let's say that the police report did not collaborate with your explanation as to what happened, then you would have to send in further documentation proving that you used your bullets to defend yourself. The next piece of documentation that you will need is court records such as a copy of the court transcript verifying that this actually did happen and you were vindicated at a court hearing.

If your story can be verified and validated, you will be given the opportunity to repurchase bullets for your gun again. If your answer to question number one was that you went shooting your gun up in the air on the Fourth of July to celebrate Independence Day, then you have used your bullets irresponsibly. Therefore, you would be denied access from repurchasing bullets again

I mean, think about it, you may have had good intentions on that day and meant nobody any harm by shooting your gun up into the air to show your patriotism.

However, can you really responsibly track those bullets that you shot up into the air? Where did they land? On top of someone's head? In someone's eye? Or maybe it came down and landed in the temple of someone's pet dog and killed it.

Tracking the Bullet will show you to be a responsible user or an irresponsible user, and if you have proven to be an irresponsible user, then you should not be allowed to purchase bullets that kill so that you do not continue this irresponsible activity possibly causing someone's death or injury.

If your answer to question number one was that you left your gun in a hiding place in your house and your son found it and shot all of the bullets out of the gun, then you have not used your bullets responsibly.

It is your responsibility as an adult to keep your gun in a safe place where it cannot be found by your son, like locking it up in a lock box with a padlock on the door.

The excuse that you were negligent in leaving it in a place that was easily accessible by your son will not bring that person back to life if that son shot himself or someone else to death.

If your answer to question number one was that your gun was stolen and that you had to buy a new gun and now you are trying to purchase bullets for this new gun, then a police report should be produce verifying that your gun was stolen. You are only allowed one time for your gun to have been stolen. If it happens again, then you will be denied access from purchasing bullets again because the person stealing your gun could be out there murdering or assaulting someone else.

So to prevent this from repeatedly happening, you will be locked out of the system for the preservation of life. It is your responsibility to keep up with your gun.

If you cannot explain what you did with your bullets and your answer to question number one is I don't know, then you are deemed irresponsible because you cannot responsibly track deadly bullets that came from your gun, or worse ended up in the hands of a deranged gangbanger's gun.

Or maybe you are supplying the criminal community such as gangbangers and drug dealers with bullets for their guns because they do not have the qualifications to purchase bullets for their guns.

In either case you are irresponsible, and you should be denied the right to purchase bullets that kill. Questions 4, 5, 6, and 7 are related to the hunting community. 4. Are you a hunter? If so, what type of hunter are you? Deer, bird, etc. 5. What type of bullets did you purchase and how many?

6a. Where did you go hunting? 6b. Do you hunt on your own land? 6c. What is your address? 7. Do you have a hunter's license to use your gun at the location where you hunt? 7a. Is it valid? 7b. What is the expiry date?

The hunting site will also have a computerized cash register with an electronic card swipe machine attached that interacts with the centralized computer database. Unless the hunting site is the person's own land where he hunts, the clerk at the computerized hunting site will also be responsible for checking your gun for the number of bullets when you arrive and when you leave. They will also have metal detectors within their facility that you will have to pass through when you arrive and when you leave.

This is for the purpose of tracking any bullets that you may have on your person. If they find any, they are to enter the total number into the tracking database computer system.

The metal detector machine will be similar to the metal detector machines that are located and used in the federal building courthouses. It should be noted that the metal detectors can come later. A clerk can just as well check for hidden bullets on the customer by having them empty their pockets and bags and shoes and socks if need be until that particular county, state, or town can afford to place metal detectors within their firing range and hunting site for accuracy of checking for hidden bullets.

More lives will be saved as a result of cutting down on carelessness, recklessness, and irresponsible behavior.

Questions 8 and 9 are for the law enforcement personnel. 8. Are you a law enforcement officer? If so, what branch? 9. Have you reported the use of your gun to your superiors?

Law officers are sworn to protect and uphold the law. They have to give an account for the use of their pistols to their superiors within their respective branches. So they are already held to a higher standard.

However, the first three questions on the application will still apply to off duty police officers as well as to the general public.

Keep in mind that the whole purpose for this recertifying application is to cut down on the gun violence in the hands of irresponsible users in areas known for gun violence, such as in metropolitan areas and big cities like Chicago, Los Angeles, and New York.

A deer hunter or farmer usually does not fall under this category if their gun usage is stellar within the confinements of the law and the location where he lives at has little to no gun crimes and he hunts deer on his own personal land. Then this hunter will naturally go through a lower degree of accountability under the Tracking the Bullet gun-control law than someone living in an area with higher gun crimes.

Police officers are held to a higher code of ethics within their respective departments, so they generally will not fall under this category while on duty.

By holding a person accountable for the way his bullets are being used by tracking his bullet usage, you now start to control the way he uses his gun in connection with safety to the public. You then start to save lives by keeping those bullets out of the hands of irresponsible users.

The gun shops and the shooting range should also be held accountable. Audits should be done on a monthly basis to ensure that the gun shops, the firing range, and the hunting site are in full compliance.

Let's say that you have an unscrupulous gun shop owner or a gun shop store clerk who is underhandedly supplying bullets to the criminal community such as gangbangers and drug dealers for profit under the table, also known as the black market.

How would you track something like this? By hiring auditors to audit the business every thirty days to ensure that he is not illegally selling bullets that kill to persons who don't qualify to purchase bullets legally. This is where additional employment will be needed by hiring more auditors.

The auditor will check the gun shop owner's receipts and records to see that they match the number that he purchased from the distributor or the manufacturer against the number of bullets that he sold legally and what he currently has left in stock. Each electronically produced report should show valid transaction on every copy of the customer receipts. There should also be a database transaction report generated that matches this data. If the figures do not match and there are discrepancies that he can't account for, such as missing bullets from his gun shop that have no valid purchase receipt, then he shall be fined $10,000 dollars for not keeping in compliance under the Tracking the Bullet gun regulating system or a considerable amount to discourage illegal behavior.

If it happens again, his business is to be closed. I mean, come on people, these are peoples' lives that we are talking about here. Is it more important to make money by allowing deadly bullets to end up in the hands of irresponsible ones than to value and save a life by regulating deadly bullet usage? The moral answer to this question would be no.

What about the firing range? Are they in compliance with their record keeping? Are they keeping accurate records of customers that enter the firing range with bullets in their guns and on their person?

What about bullets that were purchased from the firing range, can these be accounted for? Are their receipts showing that they legally sold bullets to their customers for target practice at the firing range? Do they have the appropriate metal detectors in place for scanning persons when they enter and leave the firing range? Does their inventory match what they have in stock and what they sold to the public?

Again, if there are any discrepancies that can't be accounted for, such as missing bullets from the firing range shop that are not having accurate receipts and records and no accurate records on their customers as well, then they shall be fined $10,000 dollars or a considerable amount for not being in compliance. If it happens again, then the firing range should be closed. This same rule holds true for the hunting range site as well. Again, the moral issue here is about saving lives over making a profit illegally.

As mentioned in the story in Chapter 2, the gun regulating system will create more jobs as well. Think about the additional employment that would be needed—information technology personnel to provide the computer technical skills needed to implement this; database administrators, engineers, to create, tweak, and enhance the database for optimum performance; software application developers to develop the software application to ask the appropriate questions and accept the correct answers and reject the wrong answers; computer cash register installers to install into the gun shops, the firing range, and the hunting range in various locations where these gun shops and ranges exist in all fifty states; communication line installers to communicate from the gun shop, the firing range, and the hunting range to the local police stations in all fifty states; metal detector installers for installing these devices in the firing range and the hunting range for bullet tracking purposes; additional firing range and hunting range store clerks for checking customer guns for the number of bullets found in their guns and on their person when they enter the range and leave the range for accuracy of tracking the bullets and then entering this data into the computer so that the tracking database software can track and account for these bullets; additional police station desk clerks to verify and check the accuracy of a person's documentation to show that they have used their bullets responsibly in accordance with the Tracking the Bullet to Save Lives gun regulation system; and additional auditors in all fifty states to audit and keep the gun shops, the firing range, and the hunting range honest and to discourage black market activity so that they keep in compliance with the gun regulation system called Tracking the Bullet to Save Lives. I'm sure that there are more jobs that I may have missed here, but you get the gist of what I am saying here.

Truly a win-win-win situation because not only are we implementing a superior gun-control system that can save lives without violating a person's constitutional rights to possess arms, we are also creating more jobs while doing so and boosting the economy as well.

Chapter 5

Tracking Data to Help Solve Gun Crimes

I mentioned how the Tracking the Bullet software would track how a person is using his gun. The tracking database software can be programmed to do much more. A lot of application databases offer trend analysis that can raise a red flag when something out of the ordinary starts to happen.

Let's say that under the Tracking the Bullet gun regulating system that Chicago is now well regulated. Meaning that all bullet purchases are being responsibly tracked and accounted for and the gun homicide rate has gone down tremendously due to the Tracking the Bullet gun regulation law.

Now let's say that one year the gun homicide rate starts to gradually go up in Chicago, Illinois, and the types of bullets that are now being found in the gun homicide rates are 9mm bullets. Since Chicago is well regulated under the new gun law, what is causing the gun homicide rate to go up all of a sudden?

If you recall that under the Tracking the Bullet gun regulation law there is room for allowances that certain locations could adhere to a lower accountability tracking under the Tracking the Bullet gun control law.

Now let's take as an example in the state of Illinois, such as downstate Illinois, a farm town location where there are no gun homicides and the farmers hunt on their own land. Under the Tracking the Bullet gun regulation law, this is one of the criterion for qualifying for the lower accountability.

Since the town has a lower accountability imposed under the Tracking the Bullet gun control law, they do not have to account for the bullets that they are purchasing as long as they continue to meet that criterion. However, their bullet purchase activity is still being tracked by their personal identification number or PIN.

According to the database trend analysis report within the tracking database system, a particular farmer has been purchasing huge amounts of 9mm bullets whereas before he use to purchase only a few.

Let's say that he is purchasing three hundred rounds a week. The database trend analysis would show this. Now either he is hunting and killing an awful lot of animals on his land on a weekly basis or he is supplying someone who lives in Chicago 9mm bullets.

He now becomes a possible suspect that the law enforcement agencies can focus their investigation on based on the database trend analysis. He may very well be the source to this rise in the gun homicide rate in Chicago.

Law enforcement agencies can use this data to try and see if there is a possible link between this farmer who lives in downstate Illinois on a farm and is purchasing huge amounts of 9mm bullets on a weekly basis, whereas before he only purchased about twenty a month, and the gun crime rate has increased in Chicago on a weekly basis. They can determine whether the same type of bullet that the farmer is purchasing huge amounts is now being found in the increased number of gun homicide cases in Chicago.

The Tracking the Bullet database system can at the very least help solve crimes based on the data activity of the customer's individual purchases. Again, just another plus for implementing this new Tracking the Bullet to Save Lives gun regulation law.

CHAPTER 6

What Do the Gun Laws Show Today?

While great strides have been made state by state to control the way guns are being used in the United States, obviously more is needed to compensate for the increase in gun violence and homicides. Let's take a look at the gun control laws in every state to see where the Tracking the Bullet gun regulation law can help. Regarding the state-by-state gun laws, the source of information is taken from "Wikipedia," the free encyclopedia, at the time of this writing.

Gun law in the United States is defined by a number of state and federal statutes. In the United States of America, the protection against infringement of the right to keep and possess arms is addressed in the Second Amendment to the United States Constitution. Most federal gun laws were enacted through:

- *National Firearms Act* (1934)
- *Omnibus Crime Control and Safe Streets Act of 1968* (1968)
- *Gun Control Act* (1968)
- *Firearms Owner's Protection Act* (1986)
- *Brady Handgun Violence Prevention Act* (1993)
- *Federal Assault Weapons Ban* (1994) (now defunct)

In addition to federal gun laws, most states and some local jurisdictions have additionally imposed their own firearms restrictions.

Gun laws in the United States vary from state to state and are independent of, though sometimes broader or more limited in scope than, existing federal firearms laws. Some states in the US have also created so-called assault weapon bans that are independent of, though often similar to, the expired *federal assault weapons ban*. The state-level bans vary significantly in their form, content, and level of restriction. Forty-four states have a provision in their *state constitutions* similar to the *Second Amendment* of the *Bill of Rights* (the exceptions are *California, Iowa, Maryland, Minnesota, New Jersey,* and *New York*).

Firearm license-holders are subject to the firearm laws of the state they're in, not the state in which the permit was issued. *Reciprocity* between states exists for certain licenses, such as concealed carry permits. These are recognized on a state-by-state basis. For example, Arizona recognizes a Nevada permit, but Nevada does not recognize an Arizona permit. Florida issues a license to carry both concealed weapons and firearms, but others license only the concealed carry of firearms. Some states do not recognize out of state Concealed Carry Weapon (CCW) permits at all, so it is important to understand the laws of each state when traveling with a handgun. When planning a trip it can be very confusing to match the concealed carry weapon permit to the state laws. John Thune of South Dakota introduced a national reciprocity bill, but it has never been able to advance out of Senate committees. Checking with each state's legal page is important. There are travel tools that may help shorten the search time.

Alabama

Alabama is classified as a "may issue" state; Alabama law states, "The sheriff of a county may, upon the application of any person residing in that county, issue a qualified or unlimited license to such person to carry a pistol in a vehicle or concealed on or about his person within this state for not more than one year from date of issue, if it appears that the applicant has good reason to fear injury to his person or property or has any other proper reason for carrying a pistol, and that he is a suitable person to be so licensed."

In practice, virtually all Alabama county sheriffs as of 2006 issue licenses to all "suitable persons." Application fees and other requirements such as training as well as the conduct of background checks vary from sheriff to sheriff. Alabama permits are honored in 22 states.

Alaska is the first state to adopt carry laws mimicking Vermont's (normally referred to as "Vermont Carry"), in which no license is required to carry a handgun either openly or concealed. However, licenses are still issued to residents who want them for purposes of carrying in other states via reciprocity, to be in complete compliance with Federal Gun Free School Zone act. The term "Alaska Carry" has been used to describe laws that require no license to carry handgun openly or concealed but licenses are still available for those who want them. Some city ordinances do not permit concealed carry without a concealed carry license, but these have been invalidated by the recent state preemption statute.

Arizona gun laws are found mostly in Title 13, Chapter 31 of the Arizona Revised Statutes. There is no registration or licensing of non-*NFA* firearms in Arizona. In fact, Section 13-3108, subsection B prohibits any *political subdivision* of the state from enacting any laws requiring licensing or registration. According to state law, a person must be 18 years of age to purchase any non-NFA firearm from any source; however, there is a federal age limit of 21 years on handgun purchases from federal firearms licensees. Generally, a person must be 18 years of age to possess a firearm or carry one openly, with such exceptions as are described below.

Arizona is classified as a "shall issue" state. Concealed carry permits are issued by the Concealed Weapons Permit Unit of the *Arizona Department of Public Safety*. Requirements for issuance include taking an 8-hour training class (provided by a licensed third party), submitting a finger print card, and paying a $60 fee. Applicants must be at least 21 years of age. New permits are valid for five years. Permits issued before August 12, 2005, are valid for four years. Renewing a permit requires only an application and finger print card. However, effective December 31, 2007, the finger print card requirement for renewal is scheduled to end. Arizona recognizes almost all valid out-of-state carry permits, with few exceptions.

The law regarding the carrying of firearms in motor vehicles by nonpermit holders is complex and has been further muddled by court decisions. However, it is clear that no permit is required to carry a firearm in a vehicle if the firearm is in plain view or locked in a trunk or other place not immediately accessible.

On foot, no permit is required to openly carry a firearm in a belt holster, gun case, or scabbard. Generally, a person must be at least 18 years of age to openly carry a firearm. However, this does not apply to:

- Juveniles within a private residence
- Emancipated juveniles
- Juveniles accompanied by a parent, grandparent, or guardian, or a certified hunter safety instructor or certified firearms safety instructor acting with the consent of the juvenile's parent or guardian
- Juveniles on private property owned or leased by the juvenile or the juvenile's parent, grandparent, or guardian
- Juveniles fourteen years of age and up engaged in any of the following activities:
- Lawful hunting or shooting events or marksmanship practice at established ranges or other areas where the discharge of a firearm is not prohibited
- Lawful transportation of an unloaded firearm for lawful hunting
- Lawful transportation of an unloaded firearm between the hours of 5:00 a.m. and 10:00 p.m. for shooting events or marksmanship practice at established ranges or other areas where the discharge of a firearm is not prohibited

Activities that require a firearm related to the production of crops, livestock, poultry, livestock products, poultry products, or *ratites* or in the production or storage of agricultural commodities

On September 30, 2009, a new law allowing people with concealed weapons permits to carry their guns into bars came into effect, providing they refrain from drinking alcohol in the establishments. T*he law* also allows bar and restaurant owners to post signs barring guns.

Arizona law permits the carrying of handguns by juveniles in some situations where they are prohibited by the Federal *Youth Handgun Safety Act*. However, state and local police in Arizona have shown little interest in enforcing the federal act.

The Arizona legislature has largely preempted political subdivisions (counties, cities) from passing their own firearms laws. Political subdivisions may regulate the carrying of weapons by juveniles or by their own employees or contractors when these employees or contractors are acting within their employment or contract. They may also limit the carrying of weapons to permit holders in parks that are less than one square mile in area and in public establishments and events. Public establishments and events where carry is limited to permit holders must provide secure storage for weapons onsite, which must be readily accessible upon entry and allows for immediate retrieval on exit.

Indian reservations, which comprise a large portion of the land area of the state, are exempt from the preemption statute and may have gun laws considerably more restrictive than state law. However, these laws do not usually apply to nontribal members passing through a reservation in a continuous journey on a major highway.

It is generally illegal to discharge a firearm within or into the limits of any municipality. However, this prohibition does not apply to persons discharging firearms in the following circumstances:

- On a properly supervised range
- In an area recommended as a hunting area by the Arizona game and fish department, approved and posted as required by the chief of police (Any such area maybe closed when deemed unsafe by the chief of police or the director of the Arizona Game and Fish Department.)
- For the control of nuisance wildlife by permit from the Arizona Game and Fish Department or the United States Fish and Wildlife Service
- By special permit of the chief of police of the municipality
- As required by an animal control officer in the performance of duties
- Firing blank cartridges
- More than one mile from any occupied structure

- In self-defense, or defense of another person against an animal attack if a reasonable person would believe deadly physical force against the animal is immediately necessary and reasonable under the circumstances to protect a person from harm
- In self-defense or in defense of another person against a criminal attack as permitted by the laws regarding defensive use of force

While discharging a firearm using blanks within the limits of a municipality is not specifically prohibited by law, it could still result in a *disorderly conduct* charge pursuant to *ARS 13-2904*.

State law prohibits the carrying of firearms in certain areas. These prohibited areas include:

- Hydroelectric or nuclear power generating stations
- Polling places on election day
- Secured areas of airports
- School grounds. However, this does not apply to:
  - Firearms for use on the school grounds in a program approved by the school
  - Unloaded firearms carried inside a means of transportation and under the control of an adult, provided that if the adult leaves the means of transportation, it's locked and the firearms are not visible from the outside
  - A person who is on the premises for a limited time to seek emergency aid, if such person does not buy, receive, consume, or possess alcohol while there

Game refuges. However, this does not apply to:

Persons traversing refuges or over roads therein carrying unloaded devices

- Landowners, lessees, permittees, their employees, or licensed trappers carrying arms while performing lawful duties

In addition, political subdivisions have limited power to prohibit the carrying of firearms in certain areas as described above. Carrying a firearm in a power generating station is a *felony*. Carrying a firearm in any other prohibited area, absent any other concomitant criminal conduct, is a *misdemeanor*. It is not illegal to carry a firearm in a liquor store or other establishment that sells alcohol only for consumption off the premises. State law prohibits the possession of firearms by certain categories of people. These prohibited possessors include:

- Anyone who has been found to constitute a danger to himself or to others pursuant to court order, and whose court ordered treatment has not been terminated by court order
- Anyone convicted of a felony, or who has been adjudicated delinquent for a felony, and whose State civil right to possess or carry a gun or firearm has not been restored
- Anyone who is, at the time of possession, serving a term of imprisonment in any correctional or detention facility
- Anyone who is, at the time of possession, serving a term of probation pursuant to a conviction for a domestic violence offense or a felony offense, parole, community supervision, work furlough, home arrest, or release on any other basis, or who is serving a term of probation or parole pursuant to an interstate compact
- Anyone who is an undocumented alien or a nonimmigrant alien, traveling with or without documentation for business or pleasure, or who is studying in Arizona and maintains a foreign residence, except for:
    - Nonimmigrant aliens who possess a valid hunting license or permit lawfully issued by a state in the United States
    - Nonimmigrant aliens who enter the United States to participate in a competitive target shooting event or to display firearms at a sports or hunting trade show sponsored by a national, state, or local firearms trade organization devoted to competitive or sporting use of firearms
    - Certain diplomats
    - Officials of foreign governments or distinguished foreign visitors who are designated by the United States department of state
    - Persons who have received a waiver from the United States attorney general

As is the case in many states, Arizona's prohibited possessor statute is in some ways less restrictive than the federal prohibited possessor statute found in the *Gun Control Act of 1968*, and state and local police show little interest in enforcing the federal statute.

In *Arkansas*, possession or ownership of a firearm is illegal for anyone who has been convicted of a felony, adjudicated to be mentally defective, or committed involuntarily to a mental institution.

Arkansas is a "shall issue" state for the concealed carry of firearms. Applicants must pass a background check and complete a training course to receive a new or renewal concealed carry license. An existing license is suspended or revoked if the license holder is arrested for a felony or for any violent act, becomes ineligible due to mental health treatment, or for a number of other reasons. Concealed firearms may not be carried at a courthouse, meeting place of any government entity, athletic event, tavern, or in a number of other places.

Arkansas has state preemption for most firearms laws. However, localities may enact laws regulating the discharge of firearms, or in emergency situations. Local government units and private individuals may not sue firearms manufacturers or dealers for matters relating to the lawful manufacture or distribution of firearms, except in cases of product liability or breach of contract.

Automatic weapons must be registered with the Arkansas secretary of state, in addition to being registered under federal law.

California is a "may issue" state for *concealed carry*. A license to carry a concealed firearm maybe issued or denied to qualified applicants at the discretion of the county sheriff or municipal police chief. California does not recognize concealed carry permits issued by other states.

Open carry of loaded firearms in public is generally prohibited except in unincorporated areas where the county has not made open carry illegal or where the discharge of firearms is not prohibited. Carrying of an unloaded, unconcealed firearm in plain sight is not prohibited except in areas otherwise prohibiting the carry of firearms under state or federal law, such as school zones, post offices, government buildings, state and national parks, "sterile" areas controlled by security screenings, etc.

The buyer of a firearm must fill out an application to purchase a particular gun. The firearms dealer sends the application to the California Department of Justice (DOJ), which performs a background check on the buyer. The approved application is valid for 30 days. There is a 10-day waiting period for the delivery of any firearm.

Sales of firearms from one person to another (private party transfers) must be through a licensed firearms dealer using a Private Party Transfer form. The licensed dealer may charge a $10 fee, in addition to the $25 transfer fee that the state charges. Any number of firearms maybe transferred at one time using this method. The dealer submits a Dealer's Record of Sale (DROS) form to the state, and the purchaser must wait 10 days before picking up the guns. Federally defined curio or relic long guns over 50 years old maybe sold without going through a licensed dealer.

Handgun purchases, except for private party transfers, are limited to one per 30-day period. To purchase a handgun, a buyer must have a Handgun Safety

Certificate. This is obtained by passing a written test, given by a Department of Justice certified instructor, on the safe and legal use of handguns. The certificate is valid for five years. A buyer must also perform a Safe Handling Demonstration when taking possession of a handgun. Some individuals are exempt from the Safety Certificate and Handling Demonstration requirements, including active and retired military and law enforcement personnel, hunter safety certificate holders, and concealed carry license holders.[30]

Dealers may not sell any handgun unless it is listed in the Department of Justice roster of handguns certified for sale. Listed handguns must include certain mechanical features and pass a set of laboratory tests. Private party transfers, curio/relic handguns, certain single-action revolvers, and pawn/consignment returns are exempt from this requirement.

It is illegal to sell a firearm that the state has defined as an "assault weapon," and which has been listed in the DOJ roster of prohibited firearms, which includes many military look-alike semi-automatic rifles and *.50 caliber BMG* rifles. DOJ rostered firearms maybe legally possessed if already registered with the state prior to January 2005. Military look-alike firearms that are not listed on the DOJ roster of prohibited firearms, known as "off list lowers," are legal to own and possess, as long as state laws concerning configuration are followed. It is illegal to import, sell, give, trade, or lend a *magazine* that holds more than 10 rounds of ammunition, except for fixed tubular magazines for *lever-action* rifles and *.22 caliber* rifles; however, the possession of such magazines is legal. It is illegal to possess an *automatic firearm* or a short-barreled shotgun or rifle without permission from the Department of Justice; such permission is generally not granted.[33]

On October 13, 2007, California enacted AB 1471. This controversial law requires that, effective January 1, 2010, semi-automatic handguns be equipped with microstamping technology and be listed in the roster of handguns certified for sale. When such a pistol is fired, the microstamping mechanism will imprint each cartridge case with a microscopic array of characters that will uniquely identify the gun that fired it. However, the text of this law has language that states that it will not be enforced if there is only one manufacturer that has the ability to equip this technology. As of the this writing, only one such manufacturer exists, and there are no others on the horizon.

Colorado Open carry without a permit long handguns and handguns. Technically legal in most areas unless local laws exist (City of Denver), in which case signs must be posted. Maybe interpreted as disturbing the peace by law enforcement.

Connecticut is a Shall Issue state. "Every citizen has a right to possess arms in defense of himself and the state." Article 1, Section 15.

Permits in Connecticut are first issued by the town police department, which conducts the background checks and fingerprinting. Each town is different in its willingness to approve permits, and some towns create their own requirements that go well beyond the State requirements. Meeting these town-specified requests (such as letters of reference, pictures, or an essay on why you want to have a permit to carry) does not have to be accomplished in order to get a permit. The town has 8 weeks to approve the permit. If it doesn't, the resident can appeal the ruling to the Connecticut Board of Firearms Permit Examiners, whom must grant the permit unless there is a specific reason the individual should be denied. These include:

Criminal possession of a narcotic substance; • Criminally negligent homicide; • Assault in the third degree; • Reckless endangerment in the first degree; • Unlawful restraint in the second degree; • Riot in the first degree; • Stalking in the second degree; • Has not been convicted as a delinquent for the commission of a serious juvenile offense; • Has not been discharged from custody within the preceding twenty years after having been found not guilty of a crime by reason of mental disease or defect; • Is not subject to a restraining or protective order issued by a court in a case involving the use, attempted use or threatened use of physical force against another person; • Is not subject to a firearms seizure order issued for posing a risk of personal injury to self or others after a hearing; or • Is not prohibited from possessing a firearm for having been adjudicated as a mentally incompetent under federal law.

Connecticut Residents are issued a "permit to carry pistols and revolvers," which permits both open and concealed carry[37]. Although open carry is not restricted by state law, the Connecticut Board of Firearms Permit Examiners suggests that, "every effort should be made to ensure that no gun is exposed to view or carried in a manner that would tend to alarm people who see it." Residents of other states who hold a concealed weapons permit may apply for a nonresident Connecticut permit through the mail.

Connecticut has bans on defined 'assault weapons.' However, it does not restrict magazine capacity.

Connecticut allows all NFA firearms except for selective fire machineguns. Selective fire machineguns existing in Connecticut before they were banned are grandfathered. Selective fire means that a machine gun can fire semi or fully automatic. A machine-gun that can only fire fully automatic is legal in Connecticut.

Connecticut also has a provision in the statute that if a carry permit holder loses a firearm and does not report it, they may lose the permit

Delaware there are no state permit to Purchase Long guns and handguns. There are no Firearm registration laws. There are no Assault weapon law. There are no Owner license required for long guns and handguns.

District of Columbia

In Washington, DC, all firearms must be registered with the police, by the terms of the *Firearms Control Regulations Act of 1975.*

The same law also prohibited the possession of handguns, even in private citizens' own homes, unless they were registered before 1976. However, the handgun ban was struck down by the *United States Supreme Court* in the 2008 case *District of Columbia v. Heller.* The Supreme Court ruled that the *Second Amendment* acknowledges and guarantees the right of the individual to possess and carry firearms, and therefore DC's ban on handguns was unconstitutional. A lawsuit was filed on August 8, *2009* to compel the district to issue permits to carry weapons.

Florida is a "shall issue state," and issues *concealed carry* permits to both residents and nonresidents. Florida recognizes permits from any other state that recognizes Florida's permit, provided the nonresident individual is a resident of the other state and is at least 21 years old.

Vehicle carry without a permit is allowed either in a snapped holster in plain view, or when the firearm is concealed if the firearm is "securely encased." "Securely encased" means in a glove compartment, whether or not locked; snapped in a holster; in a gun case, whether or not locked; in a zippered gun case; or in a closed box or container which requires a lid or cover to be opened for access. (Note: this legal condition is not the same as "encased securely.") Vehicle carry without a permit is permitted when concealed even if it is not "securely encased" if the firearm is not "readily accessible." Vehicle carry on one's person inside a vehicle without a permit is not allowed.

Open carry when on foot in a public area is generally not permitted, but is allowed in certain circumstances, as defined in Florida statute 790.25(3). For example, open carry is permitted while hunting, fishing, or camping, or while target shooting, or while going to or from such activities. When hunting on private land, or on properties expressly approved for hunting by the Fish and Wildlife Conservation Commission or Division of Forestry, open carry is also permitted.

State preemption laws prohibit localities from regulating firearms, other than with regards to zoning laws (i.e., for restricting where gun sellers may locate their businesses.)

Firearm regulations are uniform throughout the state, and a carry permit is valid throughout the state, in all areas other than in a few specially defined areas. These specially defined prohibited areas include:

- Federally controlled areas (such as national parks, inside the boundaries of which guns must be kept securely locked.),

- In or around specially-marked buildings/grounds (notably, mental hospitals and any hospitals with provisions to treat mental illness, where concealed carry is a felony even with a permit (FS 394.458). FS 394.458 does state concealed carry is prohibited "unless authorized by law." Since FS 790.06(12) does not prohibit concealed carry in hospitals that treat mental illness by permit holders, it can be inferred that concealed carry with a permit is allowed. Caution is advised since there currently is no case law. In other words, no case has been referred to a Grand Jury nor has any person been tried for violating the law. One Florida resident was arrested but the charges were subsequently dropped after their attorney successfully argued the permit holder was excepted. Be advised each county's prosecutor may have a different opinion.
- Any place of nuisance,
- Sheriff's Office,
- Police Station,
- Jail,
- Prison,
- Courthouse,
- Polling Place,
- Any Governmental Judicial meeting,
- Any school or college,
- Lounges,
- Bars,
- Airports,
- Professional athletic event, and
- Any federal buildings or property.

As of October 1, 2005, Florida became a "Stand-your-ground" state. The Florida law is a self-defense, self-protection law. It has four key components:

1. It establishes that law-abiding residents and visitors may legally presume the threat of bodily harm or death from anyone who breaks into a residence or occupied vehicle and may use defensive force, including deadly force, against the intruder.
2. In any other place where a person "has a right to be," that person has "no duty to retreat" if attacked and may "meet force with force, including deadly force if he or she reasonably believes it is necessary to do so to prevent death or great bodily harm to himself or herself or another to prevent the commission of a forcible felony."

3. In either case, a person using any force permitted by the law is immune from criminal prosecution or civil action and cannot be arrested unless a law enforcement agency determines there is probable cause that the force used was unlawful.
4. If a civil action is brought and the court finds the defendant to be immune based on the parameters of the law, the defendant will be awarded all costs of defense.

As of July 1, 2008, Florida became a "Take your gun to work" state. A new statewide Florida law went into effect on this date prohibiting most businesses from firing any employee with a Concealed Weapon License for keeping a legal firearm locked in their car in the company parking lot. The purpose of the new law is to allow CWL holders to exercise their Second Amendment rights during their commutes to and from work. Notable exceptions include companies with a primary business involving the manufacture of fireworks, the operating of nuclear reactors, or at sites involving Government facilities and defense contractor facilities. Currently, a test case involving *Walt Disney World Resort* is going through the courts, involving a former Disney security guard who was fired, despite having a CWL, for having a firearm locked in his car on July 1, in violation of Disney's preexisting and strict no weapons allowed policy. Disney claims that they are exempt from the new state law, on the basis of their having a fireworks license for conducting nightly fireworks shows at Disney World.

Florida law allows private firearm sales between residents without requiring any processing through an FFL. Florida law also permits larger municipalities to elect to require a concealed carry permit for a buyer to purchase a gun at a gun show from another private individual without any delay, but in practice, this applies only to a few of the largest municipalities (Miami, Orlando, etc.) where it has been invoked.

Currently, Florida's Concealed Weapon License is one of the most widely recognized, state-issued concealed weapon permit. The resident Florida Concealed Weapon License is recognized in thirty-three different states, while the nonresident Florida Concealed Weapon License is recognized in twenty-seven states.

Georgia is a "shall issue" state, and issues firearms permits to residents through a county *probate court*. Georgia recognizes permits from any other state that recognizes Georgia's permit, provided the nonresident individual would meet the eligibility requirements for a Georgia Firearms License as a resident.

Vehicle carry is allowed if the possessor is eligible for a permit. An eligible person without a permit must keep a firearm unloaded in a case separated from ammunition, or loaded and fully exposed to view (*Lindsey vs State of Georgia* indicates that the firearm must be fully visible to all possible observers), or loaded in a closed compartment of the vehicle. Permit holders may carry a firearm open or concealed anywhere within a vehicle.

State preemption laws prohibit localities from regulating the ownership, transportation, and possession of firearms. Georgia also has a law preventing localities from enacting ordinances or lawsuits to classify gun ranges as nuisances.

Firearm regulations are uniform throughout the state, and a firearms permit is valid throughout the state, in all areas other than in a few specially defined areas. These specially defined prohibited areas include:

- Federally controlled areas (such as national parks, inside the boundaries of which guns must be kept securely locked)
- Nuclear power facilities
- Any federal buildings or property
- Any public gathering (includes, but is not limited to, athletic or sporting events, churches or church functions, political rallies and/or functions, publicly owned or operated buildings)
- Any place licensed to sell alcohol for consumption on premises (excluding restaurants that serve alcohol)
- Wildlife management areas, except by a licensed hunter in an appropriate open season (not during a primitive weapon season)
- Any school building or grounds (except for authorized teachers and staff)

On May 14, 2008, Governor Sonny Perdue signed House Bill 89. The bill removed public transportation, state parks, and restaurants that serve alcohol (excluding bars) from the list of specially defined prohibited areas.

As of July 1, 2006, Georgia became a "Castle Doctrine" state, and requires no duty to retreat before using deadly force in self defense, or defense of others.

Georgia law allows private firearm sales between residents without requiring any processing through an FFL

Hawaii

Hawaii is a "may issue" state for *concealed carry*. "In an exceptional case, when an applicant shows reason to fear injury to the applicant's person or property," a license to carry a pistol or revolver maybe granted or denied at the discretion of the county police chief.[55] In practice however few if any concealed carry licenses are granted. Hawaii does not recognize concealed carry permits issued by other states.

Acquiring a firearm in Hawaii requires a permit to acquire, issued to qualified applicants by the county police chief. There is a 14-day waiting period for receiving a permit, which is then valid for 6 days. A separate permit is required for each handgun to be acquired, while a "long gun" permit can be used for any number of rifles or shotguns for a period of one year. In addition to passing a criminal background check, applicants must provide an affidavit of mental health, and agree to release their medical records. First time applicants must be fingerprinted by the FBI. When applying to acquire a handgun, a handgun safety training course affidavit or hunter's education card is also required.

Firearms acquired within the state must be registered with the chief of police within 5 days. Firearms brought in from out of state, including those owned prior to moving to Hawaii, must be registered within 3 days of arrival. Registration of firearms brought in from out of state does not involve a waiting period, however a FBI fingerprint and background check will be conducted. Registration is not required for black powder firearms or firearms manufactured before 1899.

Carrying a loaded firearm, concealed or not concealed, including in a vehicle, is a class A felony. Unloaded firearms that are secured in a gun case and are accompanied by a corresponding permit are allowed to be transported in a vehicle between the permitted owner's residence or business and: a place of repair; a target range; a licensed dealer's place of business; an organized, scheduled firearms show or exhibit; a place of formal hunter or firearm use training or instruction; or a police station.

Automatic firearms, shotguns with barrels less than eighteen inches long, and rifles with barrels less than sixteen inches long are prohibited by state law. Also banned are handgun magazines that can hold more than ten rounds of ammunition, and semi-automatic handguns with certain combinations of features that the state has defined as "assault pistols."

Idaho

Idaho is a "shall issue" state for *concealed carry*. The local county sheriff shall issue a concealed weapons permit to a qualified applicant within ninety days. Applicants must demonstrate familiarity with a firearm, generally by having taken an approved training course or by having received training in the military. A permit is valid for five years; permits issued before July 1, 2006, are valid for four years. Idaho recognizes valid concealed carry permits from any state. A concealed weapon may not be carried at a school or at a school sponsored activity, in a courthouse, in a prison or detention facility, at the Idaho State Capitol mall, or in certain other governmentally designated locations. It is unlawful to carry a concealed weapon while intoxicated.

Open carry is legal in Idaho. A concealed weapons permit is not required for open carry, nor for long guns (concealed or not). The firearm being openly carried must be clearly visible. A firearm can also be transported in a vehicle, as long as it is in plain view, or is disassembled or unloaded.

Idaho has state preemption of firearms laws, so local units of government cannot regulate the ownership, possession, or transportation of firearms. The state constitution states that "No law shall impose licensure, registration or special taxation on the ownership or possession of firearms or ammunition. Nor shall any law permit the confiscation of firearms, except those actually used in the commission of a felony."

The possession of *automatic firearms* is permitted, as long such possession is in compliance with all federal regulations

Illinois

Illinois has some of the most restrictive firearm laws in the country.

To possess or purchase firearms or ammunition, Illinois residents must have a Firearm Owner's Identification (*FOID*) card, which is issued by the state police. Generally an FOID will be granted unless the applicant has been convicted of a felony or an act of domestic violence, is the subject of an order of protection, has been convicted of assault or battery or been a patient in a mental institution within the last five years, or has been adjudicated as a mental defective. There are additional requirements for applicants under the age of twenty-one. There is no state *preemption* of firearm laws. Some municipalities, most notably *Chicago*, require that all firearms be registered with the local police department. Chicago does not allow the registration of *handguns*, which has the effect of outlawing their possession, unless they were *grandfathered in* by being registered before April 16, 1982. The Chicago suburb of *Oak Park* also has banned handguns, and *Highland Park* bars handgun possession unless the resident has obtained a permit from the police. The status of these various handgun bans has been uncertain since June 26, 2008, when the United States Supreme Court struck down Washington, DC's handgun ban in the case of *District of Columbia v. Heller*. In the months following the *Heller* decision, handgun bans were repealed in the suburbs of *Wilmette, Morton Grove, Evanston,* and *Winnetka,* but Chicago and Oak Park have fought in court to keep their current laws. The Supreme Court has agreed to review the Chicago and Oak Park handgun bans in the case of *McDonald v. Chicago*.[

Cook County has banned *assault weapons* and *magazines* that can hold more than ten rounds of ammunition. Other municipalities have also enacted various firearm restrictions. Lack of preemption makes it difficult to travel throughout Illinois with a firearm while being sure that no laws are being broken.

Illinois is one of two remaining states that have no provision for the *concealed carry* of firearms by citizens. (In compliance with the federal *Law Enforcement Officers Safety Act*, retired police officers who qualify annually under state guidelines are allowed to carry concealed.) *Open carry* is also illegal, except when in unincorporated areas where carrying is not prohibited by county law, a fixed place of business with owner's permission, or in one's abode. When a firearm is being transported, it must be unloaded and enclosed in a case. When purchasing a handgun in Illinois, there is a 72-hour waiting period after the sale before the buyer can take possession. The waiting period for *long guns* is twenty-four hours.

Indiana

Indiana has enacted state preemption of firearm laws. However, local laws passed before 1994 or for certain narrowly defined emergency situations are valid.

Indiana is a "shall issue" state for the License to carry a handgun. The Indiana license to carry allows both open and concealed carry. Most Indiana residents confuse the license to carry a handgun with a CCW. A license to carry will be issued to individuals age eighteen or older who meet a number of legal requirements. Grounds for disqualification include a conviction for a felony or for misdemeanor domestic battery. A license can also be denied if the applicant has been arrested for a violent crime and "a court has found probable cause to believe that the person committed the offense charged." Application for a license must be made to the local police department, or absent that to the county police department. It is illegal to carry a concealed weapon, even sporting arms, on school property or on a school bus, on an airplane or in the controlled section of an airport, on a riverboat gambling cruise, or at the Indiana State Fair. Indiana honors CCW licenses issued by every state (Illinois and Wisconsin do not issue CCW licenses), generally including nonresident licenses. However, Indiana residents, or nonresidents with a "regular place of business" in Indiana, must obtain an Indiana license.

Firearms dealers or private individuals may not sell any firearm to someone less than eighteen years old, or less than 23 years old if the buyer was "adjudicated a delinquent child for an act that would be a felony if committed by an adult," or to a person who is mentally incompetent or is a drug or alcohol abuser. Possession of automatic weapons by individuals or dealers who have obtained the appropriate federal license is permitted. Short-barreled shotguns are prohibited.

Indiana provides lawsuit protection to law-abiding manufacturers, sellers, and trade associations for the misuse of firearms by third parties. Lawsuits are permitted for cases of damage or injury caused by defective firearms or ammunition, or breach of contract or warranty.

Iowa

Iowa is a "may issue" state for *concealed carry*. A Permit to Carry Weapons maybe issued or denied to qualified residents at the discretion of the county sheriff, or to nonresidents at the discretion of the Commissioner of the Iowa Department of Public Safety. Applicants must successfully complete an approved training course and demonstrate proficiency with a handgun. Iowa does not recognize concealed carry permits issued by other states.

A Permit to Acquire Pistols or Revolvers is required when purchasing a *handgun* from a dealer or private party, and is obtained from the county sheriff. A Permit to Acquire shall be issued to qualified applicants aged 21 or older. It becomes valid three days after the date of application, and is valid for one year. A permit is not required when purchasing an antique handgun, which is defined as a *matchlock*, *flintlock*, or *percussion cap* pistol manufactured before 1898, or a replica of such a pistol.

Iowa has enacted state preemption of firearms laws, so local units of government may not restrict the ownership, possession, or transfer of firearms, or require their registration.

Under Iowa law, private citizens may not possess *automatic firearms*, short-barreled rifles, or short-barreled shotguns.

Kansas

Despite having relatively nonrestrictive firearms laws, Kansas remained one of the few states with no provision for the concealed carry of firearms until March 2006, when the state legislature passed Senate Bill 418, "The Personal and Family Protection Act." This bill made Kansas the 47th state to permit concealed carry in some form and the 36th state with a "shall issue" policy. The bill was passed 30-10 in the state senate and 91-33 in the state house of representatives, gaining enough votes to override a veto from Governor *Kathleen Sebelius*, who had previously vetoed several other attempts to legalize concealed carry. Under the law, the Attorney General began granting permits to qualified applicants on January 1, 2007. Previously, Kansas had allowed only *open carry* of firearms, except where prohibited by local ordinance.

On April 21, 2008, Governor Kathleen Sebelius signed a bill allowing the sale and possession of NFA weapons. The law took effect on July 1, 2008.

Kentucky has a "shall-issue" concealed carry law, set forth in *Kentucky Revised Statutes § 237.110*. Once issued, permits are valid for 5 years. Kentucky's law in this area has a few distinctive features:

- An applicant must have been a resident of Kentucky for at least 6 months before filing his or her application. A member of the U.S. military who has been stationed in Kentucky for at least 6 months at the time of filing is also eligible, regardless of his or her domicile.
- Applicants cannot be in arrears on child support obligations in an amount that would equal or exceed that accumulated in one year of nonpayment.
- The required firearms safety course can be no longer than 8 hours, and also includes a mandatory marksmanship test, in which the applicant must hit a full-sized silhouette target from 7 yards with at least 11 out of 20 rounds fired. (The distance is not stated in the statute, but is set forth in the administrative regulations governing the course.

Under KRS § 237.110 (20)(a), Kentucky recognizes all currently valid concealed carry permits issued by other U.S. jurisdictions.

Suppressors are legally transferable in Kentucky.

KRS *§ 527.020 (8)* permits firearms to be carried in a glove compartment of a motor vehicle:

A firearm or other deadly weapon shall not be deemed concealed on or about the person if it is located in a glove compartment, regularly installed in a motor vehicle by its manufacturer, regardless of whether said compartment is locked, unlocked, or does not have a locking mechanism. No person or organization, public or private, shall prohibit a person from keeping a firearm or ammunition, or both, or other deadly weapon in a glove compartment of a vehicle in accordance with the provisions of this subsection. Any attempt by a person or organization, public or private, to violate the provisions of this subsection maybe the subject of an action for appropriate relief or for damages in a Circuit Court or District Court of competent jurisdiction.

KRS *§ 237.104* prohibits the state from seizing firearms from private citizens in the event of a disaster or emergency

Louisiana

Louisiana is a "shall issue" state for *concealed carry*. The Louisiana Department of Public Safety and Corrections shall issue a concealed handgun permit to qualified applicants, after performing an NICS background check and giving the local police 10 days to provide additional information about the applicant. An applicant must demonstrate handgun proficiency by taking a training course from an approved instructor, or by having been trained while serving in the military. Concealed carry is not permitted in any portion of the permitted area of an establishment that has been granted a permit to sell alcoholic beverages for consumption on the premises, or in any place of worship, government meeting place, courthouse, police station, polling place, parade, or in certain other locations.

Louisiana has state preemption of firearms laws, except for local laws passed before July 15, 1985. Government bodies other than the state may not sue firearms manufacturers, dealers, or trade associations for damages that are the result of lawful activities. *Automatic firearms*, short-barreled *shotguns* and *rifles*, and *silencers* may not be possessed or transferred without permission of the Department of Public Safety, and must be registered with the Department.

Maine

Maine is a "shall issue" state for concealed carry. The issuing authority is the local government or local police, or the state police. A permit to carry a concealed firearm shall be issued within 30 days to a qualified applicant who has been a Maine resident for at least five years, or within 60 days to a nonresident or a resident for less than five years. The permit is valid for four years. A permit to carry a concealed firearm is required when transporting a loaded firearm, either concealed or plainly visible, in a vehicle. A permit is not required for open carry.

Maine does not currently honor concealed carry permits from other states. In 2007, the state enacted LD 148, directing the Department of Public Safety and the Attorney General to review other states' concealed weapon reciprocity agreements and actively seek reciprocity where appropriate.

Maine has state preemption of firearms laws. No political subdivision of the state may adopt any law regulating firearms or ammunition, except for regulating the discharge of firearms.

A municipality may not sue a firearm or ammunition manufacturer for damages relating to the lawful design, manufacture, marketing or sale of firearms or ammunition. A municipality may bring legal action for breach of contract or warranty.

Maryland

In addition to having a Maryland driver's license, Maryland requires all gun purchasers to watch a safety video before purchasing a "regulated firearm."

Once complete, the purchaser is issued a certification card. Every time you purchase a regulated firearm in Maryland thereafter, you must present this card to prove you have watched the safety video and understand. An online program offered by the Maryland Police Training Commission can also fill this requirement with the purchaser receiving the card at the end of the online lecture.

Maryland State Police also maintains a database of all "regulated firearms" purchased in the state. A regulated firearm in Maryland includes all handguns and so-called "assault-style" rifles. Assault style rifles include all AK-47 style rifles, CETME/G3, FAL and copies, etc. When purchasing a regulated firearm, there is a seven-day waiting period before the purchaser can pick up his gun. During this seven-day waiting period, Maryland State Police conducts its own background check, which includes juvenile records, and, if the transaction is approved, they record the firearm serial number and owner. If a legally registered gun owner commits a felony in Maryland, Maryland State Police uses their database to obtain a search warrant and confiscate any and all firearms in that person's possession.

Maryland tightly restricts the issuance and use of concealed carry permits. An individual must either prove that their employment requires the transportation of large amounts of cash or, in the case of security guards and private detectives, that the use of a handgun is required to perform his or her duties. Individuals citing personal protection must produce documentation of death threats supported by police reports; even then, permits have been denied in some cases. Almost all permits have tight restrictions on their use; for example, a licensed business owner or employee may only legally carry concealed while in the act of transporting cash from his or her business.

New handguns sold in Maryland must include a sealed envelope, provided by the manufacturer, containing a shell casing from a cartridge fired by that gun. When the gun is sold, the dealer must send the envelope with the shell casing, along with information identifying the purchaser, to the state police, for inclusion in their ballistics database, known as the Integrated Ballistics Identification System (IBIS).

Massachusetts

Massachusetts Law requires firearm owners to be licensed through their local Police Department or the Massachusetts State Police if no local licensing authority is available. A license is required by state law for buying firearms and ammunition. An applicant must have passed a State approved firearm safety course before applying for a license.

All applications, interviews, fees, and fingerprinting are done at the local Police Department then sent electronically to the Massachusetts Criminal History Board for the mandatory background checks, and processing. All approved applicants will receive their license from the issuing Police Department. All licensing information is stored by the Criminal History Board. Nonresidents who are planning on carrying in the state must apply for a temporary LTC through the State Police before their travel.

There are four different types of Firearm licenses issued in the state. The most restrictive license, the Firearm Identification License (FID), only allows the ownership of long arms (rifles or shotguns) that hold 10 rounds or less in their magazines. The Class B License To Carry (LTC) allows the ownership of handguns which hold 10 rounds or less or long arms with any capacity magazine. The Class A LTC allows the ownership of any capacity handgun, or longarm and the ability to carry concealed. Licenses are valid for 6 years.

The issuing authority official may restrict the Class A LTC as he wishes, such as target and hunting only, which may affect the authorization of an individual to carry concealed.

The final license is only issued for "machine guns."

A license to possess or carry a machine gun maybe issued only to a firearm instructor certified by the Criminal Justice Training Council for the sole purpose of firearm instruction to police personnel, or to a bona fide collector of firearms upon application or renewal of such license.

A "bona fide collector of firearms," for the purpose of issuance of a machine gun license, shall be defined as an individual who acquires firearms for such lawful purposes as historical significance, display, research, lecturing, demonstration, test firing, investment or other like purpose.

For the purpose of issuance of a machine gun license, the acquisition of firearms for sporting use or for use as an offensive or defensive weapon shall not qualify an applicant as a bona fide collector of firearms.

All private sales are required to be registered through an FA-10 form with the Criminal History Board, Firearm Records division. The state has an assault weapons ban similar to the expired Federal ban. Massachusetts is a "may issue" state and all classes of LTCs, are issued in a highly discretionary manner.

FIDs are "Shall issue," except if the applicant fails a background check. Massachusetts law does not recognize the *Firearm Owners Protection Act*. State law requires all firearms to be stored in a locked container, or with a *trigger lock*. If in a vehicle, firearm must be carried in the trunk of the vehicle in a locked container, unless the licensee has a Class A unrestricted license, in which case the firearm must be under his direct control. Any firearms that are found to be unsecured maybe confiscated by law enforcement officers and license maybe revoked. In the event a license is revoked for any reason, law enforcement will confiscate all weapons and store for 1 year before destroying them unless the revoked licensee transfers ownership to a properly licensed party who then claims the firearms

Michigan

The State of Michigan claims complete preemption of laws in regard to ownership and the carrying of firearms. Generally, federal, state and local law enforcement agencies, and agents thereof acting in an official capacity, are exempt from Michigan's firearms regulations. The Constitution of the State of Michigan of 1963[121] Article 1, Section 6 reads:

Every person has a right to keep and possess arms for the defense of himself and the state.

A complete listing of Michigan's firearms laws can be found in the publication "Firearms Laws of Michigan"

Firearm Defined

The word "firearm," except as otherwise specifically defined in the statutes, shall be construed to include any weapon from which a dangerous projectile maybe propelled by using explosives, gas or air as a means of propulsion, except any smooth bore rifle or handgun designed and manufactured exclusively for propelling BB's not exceeding .177 calibre by means of spring, gas or air.

Purchasing Firearms in Michigan

In *Michigan*, rifles and shotguns maybe purchased by anyone 18 or over who is not subject to restrictions that are usually based on criminal or mental health history, no licensing or permit to purchase is required for long guns. A Permit to Purchase is required to purchase a handgun, anyone purchasing a handgun must be 21 years of age or more to purchase from a federally licensed firearms dealer due to federal law; however, in the case of a private sale the purchaser need only be 18 or older.

In Michigan, a person "shall not purchase, carry, or transport a pistol in this state without first having obtained a license for the pistol," as prescribed in MCL 28.422.

An individual must apply to their local police or sheriff's department for a License to Purchase a Pistol prior to obtaining a pistol. A license to purchase is not needed for an individual with a CCW license. However, a NICS check must be completed by the FFL (Federal Firearms Licensee) prior to the transfer of the firearm. The police authority will check for any criminal record at both the state and national level. The applicant must answer gun-related questions on a Basic Pistol Safety Questionnaire, with at least 70 percent correct, and swear before a notary that they meet the statutory requirements to own a pistol. The License to Purchase a Pistol form must be completed even though the applicant may already have possession of a pistol, such as through an inheritance. Federal firearms licensed dealers are not exempt from this section of the law and must also get a license any time they purchase/acquire a pistol from an individual or another gun dealer. There is an exemption only for dealers purchasing pistols directly from the manufacturer or wholesaler. A License to Purchase a Pistol is valid for 10 days to purchase a pistol. The seller must sign the license and keep one copy for his/her records. An individual must return to the local police department within 10 days of purchasing the pistol, return the two remaining copies of the license, and present the pistol for a Safety Inspection Certificate. Dealers are exempt from the safety inspection requirements on pistols kept solely for the purpose of resale. Some agencies require all unused license to purchase forms be returned to them for record keeping purposes. These forms are licenses to purchase a pistol and the purpose is not to circumvent the required NICS (National Instant Check System) check when buying a shotgun or rifle from an FFL dealer.

Concealed Carry in Michigan

Michigan's concealed carry law is "shall issue," meaning that anyone over 21 may obtain a license to carry a concealed handgun if they are not prohibited from owning firearms, have not been found guilty or been accused of certain felonies or misdemeanors within a certain time period, and have completed state approved firearms training. Concealed Pistol License (CPL) holders are exempt from the obtaining a License to Purchase a Handgun; however, they must fulfill the registration requirement.

Individuals licensed to carry a concealed pistol by Michigan or another state are prohibited from carrying a concealed pistol on the following premises: Schools or school property, public or private day care center, public or private child caring agency, or public or private child placing agency, sports arena or stadium, a tavern where the primary source of income is the sale of alcoholic liquor by the glass consumed on the premises, any property or facility owned or operated by a church, synagogue, mosque, temple, or other place of worship, unless the presiding official allows concealed weapons, an entertainment facility that the individual knows or should know has a seating capacity of 2,500 or more, a hospital, a dormitory or classroom of a community college, college, or university, and casinos. "Premises" does not include the parking areas of the places listed above, excluding casino parking. A pistol is subject to immediate seizure if the CPL permit holder is carrying a pistol in a "pistol free" area.

An individual licensed to carry a concealed pistol who is stopped by a police officer (traffic stop or otherwise) while in possession of a pistol shall immediately disclose to the police officer that he or she is carrying a concealed pistol either on their person or in their motor vehicle.

On March 29, 2001, per Administrative Order 2001-1 of the Michigan Supreme Court: "Weapons are not permitted in any courtroom, office, or other space used for official court business or by judicial employees unless the chief judge or other person designated by the chief judge has given prior approval consistent with the court's written policy."

Open carry is legal in Michigan, but choosing to do so in populated areas may result in being charged with disturbing the peace or even brandishing. However then-Attorney General *Jennifer Granholm* released an opinion that open carrying is neither brandishing nor disturbing the peace.

Michigan formerly did not allow ownership of NFA firearms, though Attorney General *Mike Cox* has written an Attorney General's Opinion that allows for machine guns to be legally transferable to Michigan residents who comply with federal laws. Suppressors (silencers), however, are still illegal and nontransferable in Michigan.

Michigan prohibits the possession of Tasers or stun guns by private citizens, regardless of CPL status

Minnesota State Permit to Purchase handguns is required. State Permit to Purchase long guns is not required. Relevant statutes §624.7131 Notes: Permit to purchase required to transfer handguns through FFL's and for "military-style assault weapons." Minnesota does not require concealment when carrying. With a carry permit, one may carry openly in accordance with the carry law. Minnesota State Statute 624.714

Mississippi

Mississippi is a "shall issue" state for concealed carry. The Mississippi Department of Public Safety shall issue a license to carry a concealed pistol or revolver to a qualified applicant within 120 days. The license is valid for four years. Concealed carry is not allowed in a school, courthouse, police station, detention facility, government meeting place, polling place, establishment primarily devoted to dispensing alcoholic beverages, athletic event, parade or demonstration for which a permit is required, passenger terminal of an airport, "place of nuisance" as defined in Mississippi Code section 95-3-1, or a location where a sign is posted and clearly visible from at least ten feet away saying that the "carrying of a pistol or revolver is prohibited." A license to carry a concealed pistol or revolver is required for open carry. A license is not required for transporting a concealed or visible firearm in a vehicle.

Mississippi has state preemption of many but not all firearm laws. No county or municipality may adopt any ordinance that restricts or requires the possession, transportation, sale, transfer or ownership of firearms or ammunition or their components. However, local governments may regulate the discharge of firearms, the carrying of firearms at a public park or public meeting, or the use of firearms in cases of insurrection, riots and natural disasters.

Lawsuits against manufacturers, distributors, or dealers for damages resulting from the lawful design, manufacture, distribution or sale of firearms are reserved to the state. However, local governments may bring suit for breach of contract or warranty or for defects in materials or workmanship.

Missouri

Peaceable journey and RV law

Missouri has a "peaceable journey" under Missouri Statutes 571.030 which law says it is not illegal to carry the weapon in a passenger compartment of a vehicle as long as (1) the concealable firearm is otherwise lawfully possessed, (2) the person is 21 or older, or (3) the person is in his dwelling unit (e.g., RV) or upon premises over which the person has possession/authority/control, or is traveling in a continuous journey peaceably through this state.

The same applies (it is not a crime) when the person is 21 and possesses an exposed firearm for the lawful pursuit of game.

Open Carry

Missouri does allow open carry of firearms for those age 21 or older. However, city, county, and municipalities are allowed to pass laws and ordinances restricting this. It is advisable to check local laws and ordinances before openly carrying a firearm within Missouri.

Concealed Carry

Missouri Statute 571.070 (8/28/2007) says that unlawful possession of any firearm (including concealable firearms) is a class C felony.

Missouri Statute 571.121 (8/28/2007) says (a) you have to carry permit with you when you carry the concealed weapon and if you don't have it with you, it's not a crime, but you can be fined up to $35, and (b) director of revenue issues a driver's license or a state ID with a CCW endorsement that reflects that you can carry concealed (and they cannot be held responsible for what you do.)

Montana

Montana has some of the most permissive gun laws in the United States. It is unique in having no state level prohibited possessor statute, although the state preemption statute allows local governments to prohibit firearms possession among felons and mental incompetents.[

Montana is a "shall issue" state for concealed carry. The county sheriff shall issue a concealed weapons permit to a qualified applicant within 60 days. Concealed carry is not allowed in government buildings, financial institutions, or any place where alcoholic beverages are served. Carrying a concealed weapon while intoxicated is prohibited. No weapons, concealed or otherwise, are allowed in school buildings. Montana recognizes concealed carry permits issued by most but not all other states. Concealed carry without a permit is generally allowed outside city, town, or logging camp limits. Transporting a firearm in a motor vehicle is legal. Open carry is also generally permitted.

Montana has state preemption of most firearms laws. Local units of government may not prohibit, register, tax, license, or regulate the purchase, sale or other transfer, ownership, possession, transportation, use, or unconcealed carrying of any weapon. However, local governments may restrict the firing of guns, or the carrying of firearms at public assemblies or in public buildings or parks.

Montana has a number of restrictions on lawsuits against firearms manufacturers, dealers, or trade associations. Such lawsuits maybe filed by the state, but not by local governments.

Montana House Bill 246, the *Montana Firearms Freedom Act*, was signed into law by Governor *Brian Schweitzer* on April 15, 2009, and will become effective October 1, 2009. This legislation declares that certain firearms and firearms accessories manufactured, sold, and kept within the state of Montana are exempt from federal firearms laws, since they cannot be regulated as interstate commerce.

Nebraska

In *Nebraska*, to purchase a handgun, a permit to purchase is required. Rifles and shotguns are not subject to gun laws more restrictive than those at the federal level. As of January 1, 2007, shall issue concealed handgun permits (CHPs) are being issued by the Nebraska state police. NFA firearms (machine guns, short barreled shotguns, short barreled rifles, and silencers) are legal to own as long as they are compliant with federal law.

In *Lincoln*, municipal code section 9.36.100 prohibits the possession of a firearm by anyone who has been convicted of certain misdemeanors within the last ten years. These include stalking, violation of an order of protection, impersonating a police office, and public indecency.[148] The Lancaster County Sheriff will not issue a Nebraska permit to purchase a handgun if the applicant is a Lincoln resident and is prohibited by this law from possessing firearms.

Nevada

Nevada law does not require the registration of firearms. However, handgun owners in *Clark County* must register their concealable firearms at a law enforcement agency within an incorporated city of Clark County.

Nevada is a "shall issue" state for concealed carry. The county sheriff shall issue a concealed firearms permit to qualified applicants. A person must take a class to receive the CCW concealed carry permit and must qualify by demonstrating use of the exact model handgun that the person will carry.

Nevada has an open carry law that permits a person to carry a handgun in plain view, however, there are exceptions. For instance, Clark County requires you to have a registration to open carry, and you must carry your registration with you when you are carrying a gun.

States that honor Nevada's CCW permit: Alaska, Arizona, Florida*, Idaho, Indiana, Kansas*, Kentucky, Louisiana, Michigan*, Minnesota, Missouri, Montana, Nevada, Oklahoma, South Dakota, Tennessee, Texas, Utah (*Residential Permits Only)

State CCW permits that Nevada honors: Alaska, Arkansas, Kansas, Louisiana, Michigan, Missouri, Nebraska, Nevada, Ohio, Tennessee

Nevada is a traditional open carry state with seemingly complete state preemption of firearms laws. However, several localities have passed and are illegally enforcing "Deadly Weapons" laws which conflict with the preemption laws. Were this not the case, Nevada would qualify as a "Gold Star" open carry state. Effective October 1, 2007, is legislation that prohibits counties/cities/towns from enacting ordinances more restrictive than state law-the legislature reserves for itself the right to legislate firearms law. This law is retroactive. Hence the more restrictive ordinances in North Las Vegas and Boulder City are null and void.

Also effective October 1, 2007, a CCW permittee can qualify with a revolver and thereafter carry ANY revolver (or derringer)-the permit will simply state "Revolvers Authorized"; however, one must continue to qualify with each make/model/caliber of semi auto pistols.

The CCW permit costs $105, and is valid for 5 years for residents and 3 years for nonresidents.

New Hampshire

New Hampshire does not require a license for Open Carry, but carry of a loaded pistol or revolver in a motor vehicle, or concealed, does require a license. Note that the NH license is issued for carry of a "pistol or revolver," and is not a license to carry "weapons" as exists in some other states. The NH license is issued by the local police dept at a cost of $10 for residents, and by the NH State Police at a cost of $100 for nonresidents. The term of issue of the license is four years for nonresidents, and at least four years for residents. Turnaround time is generally 1-2 weeks, with fourteen days being the maximum time allowed by law.

New Jersey

In *New Jersey*, firearm owners are required to get a lifetime Firearm Purchaser card for the purchase of rifles, shotguns or handguns as well as ammunition. To purchase a handgun, a separate permit is needed from the local police department for each handgun to be purchased and expires after ninety days. NJ law says that the handgun purchase permit must be issued within 30 days, but it is not uncommon for it to take several months to be issued. Capacities of semiautomatic handguns and rifles (total in magazine excluding chamber) are limited to 15 rounds or less. New Jersey also prohibits the purchase, use, and possession of *hollowpoint* ammunition in most circumstances. New Jersey has its own ban on various semi-automatic firearms as well. Police are bound by the Assault Weapons Ban (AWB) in NJ and cannot own those firearms unless they are signed off by the Chief as used in an official capacity. They are exempt from the magazine limits when used in a duty/off duty firearm and approved by the department.

It is the express policy of New Jersey legislative and law enforcement authorities that the carrying of a handgun on one's person be strictly limited only to those earning a living through the carrying of a handgun. Theoretically, there exists a route for a civilian to legally carry a handgun, but in reality, is practically nonexistent.

Effective January 1, 2010, New Jersey will limit handgun purchases to one per 30-day period.

New Mexico

New Mexico laws governing the possession and use of firearms include those in New Mexico Statutes Chapter 30, Article 7, "Weapons and Explosives."

New Mexico has state preemption of firearms laws, so local governments may not restrict the possession or use of firearms. In 1986, Article 2, Section 6 of the state constitution was amended to say, "No law shall abridge the right of the citizen to keep and bear arms for security and defense, for lawful hunting and recreational use and for other lawful purposes, but nothing herein shall be held to permit the carrying of concealed weapons. No municipality or county shall regulate, in any way, an incident of the right to keep and bear arms."

New Mexico is a "shall issue" state for the concealed carry of handguns. An applicant for a concealed carry permit must be a resident of New Mexico and at least 21 years of age. Each permit specifies the category and caliber of handgun that maybe carried, but is also valid for a smaller caliber. The applicant must complete a state approved training course that includes at least 15 hours of classroom and firing range time, and must pass a shooting proficiency test for that category and caliber of handgun. A permit is valid for four years, but license holders must pass the shooting proficiency test every two years.

New Mexico recognizes concealed carry permits issued by 20 other states: Alaska, Arizona, Colorado, Delaware, Florida, Kentucky, Michigan, Minnesota, Missouri, Montana, North Carolina, North Dakota, Ohio, Oklahoma, South Carolina, Tennessee, Texas, Utah, Virginia, and Wyoming.

Even with a concealed carry permit, it is not legal to carry a firearm into a federal building, state building, school, or restaurant that serves alcohol. However, carrying of a licensed concealed weapon into a store that sells alcohol for offsite consumption (i.e., Grocery store, gas station, liquor store) is legal (note that "open carry" is expressly disallowed in this case).

New Mexico has an "extended domain" law, which means that a person's vehicle is considered an extension of their home. It is therefore legal to carry a loaded firearm without a permit, openly or concealed, anywhere in a vehicle. On foot, no permit is required to carry a firearm unless it is both loaded and concealed.

Concealed carry of an unloaded firearm is legal without a permit in New Mexico, except in establishments that sell alcohol.

New York

New York State, by many measures, is practically the strictest state in the nation as far as the procedure for obtaining a handgun (pistol) carry license is concerned. However, unlike many of the other extremely strict states (such as New Jersey and Illinois), once a New York State pistol license is obtained (with varying degrees of difficulty depending on the county you live in), the restrictions on carrying handguns vary greatly from pro-gun jurisdictions to anti-gun jurisdictions.

Handgun possession in New York State is strictly limited to allow only those individuals who hold a current, valid, handgun license (pistol license) issued by a jurisdiction (county or major city) within New York State to purchase, possess or carry a handgun. (NY Penal law 265.01) Pistol licenses are not issued to out-of-state residents (or even part-time residents), and no licensing reciprocity agreements with any other states exist. There are no provisions whatsoever for an out-of-state handgun owner (other than law enforcement/military) to bring a handgun to, and carry it in, New York State. Some states will recognize NY's handgun license.

Simply traveling through New York State while in possession of a handgun, for any purpose, without a New York State pistol license, is legally risky. New York State law does include a very limited exception for passing through the state for target competition purposes, but the language is exceptionally strict. Traveling through New York City, even with a license issued from another jurisdiction within New York State, must be done carefully (locked box, in vehicle's trunk, no unnecessary stops).

Application for a handgun license is through the individual's county (or major city) of primary residence, usually the police/sheriff's department, or a separate licensing authority (i.e. the "Pistol Clerk"). After initial approval on the county level, the application is then passed on to the New York State Police for further approval. The applicant will be required to ask close friends or associates to act as personal references, these individuals may be required to fill out forms, that vary in length by county, attesting to the applicant's good character. Pistol licenses can take from less than four months for approval to more than six months, but NY law only allows six months to process a license. There is no "shall-issue" provision in New York State pistol licensing law.

Pistol licenses are generally of two types, carry or premises-only (most common in New York City) issued under NY Penal Law 400. Restrictions can be placed on either type of license, for example, a number of jurisdictions allow handgun license holders to carry handguns only while in the field hunting (sportsman's license) and/or traveling to and from the range (target license). These restrictions, however, are administrative in nature; carrying a licensed, registered handgun outside of the restrictions indicated on the license should result in administrative (suspension, revocation) penalties only.

All handguns possessed within New York State (except antiques or replicas of antiques) must be registered, with each handgun's registration indicated on the owner's pistol license. All handguns, including antiques and replicas, must be registered in order to be legally loaded and fired. Some counties limit who can register a handgun on their license with some allowing cross registration of a handgun from anyone to family members only to no handgun can be cross registered. NY law does not address this issue. Sharing use of a handgun not listed on your license is only allowed at a certified range with the licensed handgun owner being present. (See NY PL 265.20 7-a) A pistol license is required to physically examine a handgun for purchase. A separate purchase document is required for each handgun purchase that is obtained by filing an amendment with the local authority.

In addition to laws pertaining to the entire state of New York, there are additional laws and statutes pertaining to licensing and permits in some of the major cities of the state; any city with a population of over 100,000 is allowed to pass additional laws. Cities with stricter laws include Albany, Buffalo, Rochester, and *New York City*.

The cost and renewal of handgun licenses vary from county to county. Importantly, concealed carry handgun licenses issued in New York City are valid in the rest of the state however premises-only licenses are not valid in the rest of the state. All Concealed carry licenses issued outside of New York City are valid throughout New York State except New York City.

Restrictions on New York State handgun licenses vary wildly from jurisdiction to jurisdiction. For example, it is practically impossible to be issued a carry pistol license in New York City, unless the license applicant is a celebrity or employed in the security industry. Most licenses issued in New York City are for on-premises possession only, carrying to and from the range must utilize a "locked-box." Periodic renewal fees, even on restricted carry licenses, are highly prohibitive as well. Nassau, Suffolk, Westchester and several other suburban counties are only slightly less prohibitive, allowing a highly restricted "to and from the range only" form of concealed carry.

In contrast to "practically no carry" New York City, and some county Judges who only issue "to and from target shooting, hunting and fishing" licenses, many upstate New York counties issue "unrestricted" handgun licenses that allow unrestricted concealed carry of a loaded handgun (except for important exceptions such as schools, court houses/rooms, secure areas of airports). Some of the most rural upstate counties (such as Delaware County) specifically do not enforce the "concealed" language in New York State's licensing law, thereby effectively allowing open carry.

Paradoxically, except for visiting New York City (which effectively invalidates any carry license), the restrictions (or lack thereof) as they appear on the license stay with the license as the individual travels from region to region within the state. For example, the holder of a Delaware County pistol license (unrestricted carry) can carry his handgun into a diner in Suffolk County, while his Suffolk County friend cannot. Most counties in NY issue "lifetime" licenses: *PL400.00-10. License: expiration, certification and renewal. Elsewhere than in the City of New York and the counties of Nassau, Suffolk and Westchester, any license to carry or possess a pistol or revolver, . . . , shall be in force and effect until revoked as herein provided.* Renewable licenses last from the three year

NY City's license to five years in other counties with NY Cities license costing $340 every three years.

This striking dichotomy in New York State handgun license policies (upstate rural/downstate urban) is an outgrowth of three specific cultural forces; the strength of home rule in New York State, the prevalence of conservative political forces upstate, as well as the gun culture during the various hunting seasons in the rural counties. Not all of the most pro-gun counties of New York are particularly far from New York City either, many a tourist getting away from New York City for a weekend trip to the country has been quite surprised at the prevalence of openly carried firearms of all types only several hours from home.

Rifles and shotguns do not have to be registered in any jurisdiction within New York State except for New York City, which requires registration. Laws pertaining to the handling of rifles and shotguns are in sharp contrast to those of handguns. For example, licensed carry of a handgun on one's person allows the handgun to be fully loaded, including within an automobile, while visiting a place of business or while crossing a public road while hunting. A rifle or shotgun cannot be kept loaded in any of the above circumstances except for a self-defense emergency. A range officer would not normally take exception to a target shooter driving to the range and entering the parking lot carrying a licensed, loaded, holstered handgun; doing so with a loaded rifle or shotgun would cause quite a stir.

Most handgun licenses are issued under 3 sections of NY PENAL law 400 Section 2:

(a) have and possess in his dwelling by a householder; (b) have and possess in his place of business by a merchant or storekeeper; . . . (f) *have and carry concealed*, without regard to employment or place of possession, by any person when proper cause exists for the issuance thereof;

Section 2 f (above) is the section of NY law that most people are issued a handgun license under including the made up, "Sportsman's license" that some counties call their restricted license. This section only allows "have and carry concealed" so open carry in NY State is not allowed however some police departments in some counties may choose to ignore someone who does open carry.

NY *Penal* law: 265.20 section 7-e allows for youth between 14 and 21, inclusive, to shoot a handgun at a range as long as several simple requirements are met.

New York State is a particularly interesting case, because New York separates all of *New England* from the bulk of the United States. This means that under the *Firearm Owners Protection Act*, all people traveling through New York City and New York State with firearms must have them unloaded and locked in a hard case where they are not readily accessible (e.g., in the trunk of a vehicle) and can never be in possession of a high capacity feeding device made post ban.

New York State has a ban that is an almost exact mirror of the *Federal Assault Weapons Ban*, except that it does not have a *sunset provision*.

Most of New York State gun laws are covered in two sections of New York Penal law. Article 265-(265.00-265.40) *firearms and other dangerous weapons*; Weapons Crimes, Firearms and Other Dangerous Weapons, list definitions and legal violations. This sections includes the banning of possession of a handgun, ("firearm" under definition 3,) by anyone in New York State. Section 265.20 includes exemptions to the handgun ban including to those who have a license issued under Article 400-(400.00-400.10) *licensing and other provisions relating to firearms*; Licenses to Carry, Possess, Hunting and Target, Repair and Dispose of Firearms.

New York State requires that anyone buying a gun at a gun show must have a background check done.

North Carolina

To acquire a handgun in *North Carolina* (including private sales, gifts, and inheritance) an individual must go to the county sheriff's office in the county in which they reside and obtain a pistol purchase permit. This is not required if one has a CCW (*Carrying a Concealed Weapon*) permit. State law requires the applicant to appear in person with government ID, pay a $5 fee, undergo a background check similar in scope and scrutiny to *NICS*, and have a reason for owning a pistol (hunting, target shooting, self defense, or collecting). Because there are 100 different county sheriffs in North Carolina, there are different sets of rules and requirements for obtaining such a permit, which can be determined arbitrarily by the local sheriff. Some sheriffs impose other restrictions such as a limit on the number of permits applied for at a time, waiting periods, and/or proof of good moral character (a witness or references, in some cases notarized with affidavits). This requirement is a holdover from *Jim Crow laws* that were designed to prevent African Americans and other minorities from obtaining handguns.[173]

Durham County requires the registration of handguns. In accordance to North Carolina Law, no other county or local government may require handgun registration.

North Carolina is a "shall issue" state for the concealed carry of handguns. Application for a concealed carry license is made through the local county sheriff's office. Applicants must complete a state approved training course. A CCW license is valid for a period of five years. Regardless of the possession of a CCW permit, absolutely no person may possess a concealed weapon at any government-run facility or any educational establishment.

North Carolina honors concealed carry permits issued by Alabama, Alaska, Arizona, Arkansas, Colorado, Delaware, Florida, Georgia, Idaho, Indiana, Kansas, Kentucky, Louisiana, Michigan, Mississippi, Missouri, Montana, New Hampshire, New Mexico, North Carolina, North Dakota, Ohio, Oklahoma, Pennsylvania, South Carolina, South Dakota, Tennessee, Texas, Utah, Virginia, Washington, and West Virginia. North Carolina's permit is valid in approximately thirty states, more than any other CCW permit.

Open Carry is also legal throughout North Carolina except within the town of *Cary*, which forbids it by local ordinance. In the city of *Chapel Hill*, open carry is restricted to guns of a certain minimum size, under the theory that small, concealable weapons are more often associated with criminal activity. No permit is required to carry a weapon openly in North Carolina.

North Dakota

North Dakota is a "shall issue" state for concealed carry. The North Dakota Bureau of Criminal Investigation (BCI) shall issue a concealed weapon permit to a qualified applicant. The applicant must pass a written exam and submit an application to the local law enforcement agency, which conducts a local background check before forwarding the application to the BCI. The permit is valid for three years. A concealed weapon permit is required when transporting a loaded firearm in a vehicle. Concealed carry is not allowed in an establishment that sells alcoholic beverages or in a gaming site. Concealed carry is also not allowed, unless permitted by local law, at a school, church, sporting event, concert, political rally, or public building.

North Dakota has state preemption of firearms laws. No political subdivision may enact any ordinance relating to the purchase, sale, ownership, transfer of ownership, registration, or licensure of firearms and ammunition which is more restrictive than state law.

Firearms manufacturers, distributors, and sellers are not liable for any injury suffered because of the use of a firearm by another. However, they may be sued for breach of contract or warranty, or for defects or negligence in design or manufacture.

Ohio

In April 2004, *Ohio's* concealed carry statute went into effect. The law (Ohio Revised Code 2923.12, et seq.) allows persons 21 and older to receive a concealed handgun license provided that they receive a minimum of 12 hours of handgun training (10 hours of classroom instruction and 2 hours of range time) from a certified instructor, demonstrate competency with a handgun through written and shooting tests, pass a criminal background check, and meet certain residency requirements.

The licenses are issued by county sheriffs.

The statute prohibits any person with any drug conviction from receiving a license, as well as any person convicted of a felony and those who have been convicted of certain misdemeanor crimes of violence within three years (ORC 2923.125).

The law contains language that asserts it is a "law of general application" and thus supersedes any local ordinances that are more restrictive than state law. However, as of July 2006, at least two court cases brought by municipalities are challenging this language as being in violation of Ohio's Constitution, both of which have been denied by appeals courts as having no "merit" and being in direct violation of Section 3, Article XVIII of the Ohio Constitution. Often called the "home rule" amendment, Section 3, Article XVIII states that municipalities "shall have authority to exercise all powers of local self-government and to adopt and enforce within their limits such local police, sanitary and other similar regulations, as are not in conflict with general laws." A law is a general law if it complies with the following:

1. What is a General Law? To constitute a general law for purposes of the Home Rule Amendment, a statute must (1) be part of a statewide and comprehensive legislative enactment (2) apply to all parts of the state alike and operate uniformly throughout the state (3) set forth police, sanitary, or similar regulations, rather than merely grant or limit legislative power of a municipal corporation to set forth police, sanitary, or similar regulations (4) prescribe a rule of conduct upon citizens generally. City of Canton v. State (2002), 95 Ohio St. 3d 149, 766 NE2d 96

Courts have found that the new CCW rules fall under all of the above requirements.

Ohio's concealed handgun law allows for reciprocity with other states with "substantially comparable" statutes, and to date Ohio has reciprocity with 21 other states. States that honor this state's Permit/License are listed below with 26 more in discussions about reciprocity:

- Alaska
- Updated: 2007-12-20
- *http://www.dps.alaska.gov/Statewide/PermitsLicensing/reciprocity.aspx*
- Arizona
- Updated: 2006-12-24
- *http://www.azdps.gov/ccw/reciprocity/default.asp*

- Nonresident permit/license OK
- Delaware
- Updated: 2005-06-02
- *http://attorneygeneral.delaware.gov/crime/concealedweapons.shtml*
- Florida
- Updated: 2005-06-01
- *http://licgweb.doacs.state.fl.us/news/concealed_carry.html*
- Resident permit/license only
- Idaho
- Updated: 2005-06-01
- *http://www.isp.state.id.us/patrol/faqs.html*
- Indiana
- Updated: 2004-01-09
- *http://www.in.gov/isp/faq/index.html*
- Kentucky
- Updated: 2004-01-09
- *http://www.kentuckystatepolice.org/conceal.htm*
- Michigan
- Updated: 2005-06-01
- *http://www.michigan.gov/ag/0,1607,7-164-17334_17362_22672-60639—00.html*
- Resident permit/license only
- Missouri
- Updated: 2004-02-27
- Montana
- Updated: 2005-06-01
- *http://www.doj.mt.gov/enforcement/criminaljustice/concealedweapons.asp*
- New Mexico
- Updated: 2005-11-30
- Nonresident permit/license OK
- North Carolina
- Updated: 2005-08-21
- *http://www.ncdoj.com/law_enforcement/cle_handguns.jsp*
- Oklahoma
- Updated: 2005-06-02
- *http://www.oag.ok.gov/*
- South Carolina
- Updated: 2006-12-27
- *http://www.ag.state.oh.us/le/prevention/concealcarry/reciprocity.asp*
- Resident permit/license only

- South Dakota
- Updated: 2005-07-01
- *http://www.sdsos.gov/adminservices/concealedpistolpermits.shtm*
- Nonresident permit/license OK
- Tennessee
- Updated: 2005-06-02
- *http://www.tennessee.gov/safety/handgun/reciprocity.htm*
- Utah
- Updated: 2009-06-03
- *http://publicsafety.utah.gov/bci/FAQother.html*
- Vermont
- Updated: 2004-09-11
- See general notes section of Vermont page
- Virginia
- Updated: 2004-07-30
- *http://www.vsp.state.va.us/cjis_reciprocity.htm*
- Washington
- Updated: 2004-06-30
- *http://www.atg.wa.gov/firearms/states.shtml*
- Wyoming
- Updated: 2004-07-19
- *http://attorneygeneral.state.wy.us/dci/CWP.html*

An Ohio CCW license does not allow totally unfettered carry. Any owner of private property can ban concealed handguns by posting a sign in clear view, and most government buildings are off-limits as well as hospitals and schools and most religious places as long as they are clearly marked (to be clearly marked, you *must* have a sign clearly posted by your entrances). Ohio statute ORC 2923.16 allows for three ways for a licensee to carry a concealed handgun in a motor vehicle (which includes motorcycles):

- In a closed case, bag, box, or other container that is in plain sight and that has a lid, a cover, or a closing mechanism with a zipper, snap, or buckle, which lid, cover or closing mechanism must be opened for a person to gain access to the handgun;[185]
- In a closed glove compartment or console, or in a case that is locked;
- In a holster secured on the person.
- This is a change from the earlier version of the bill that required the weapon to be in a locked box or in plain sight and secured to the person.

On 2006-11-29, the Ohio legislature approved Amended Substitute House Bill 347. This bill would preempt all firearms regulation, thus removing any doubt as to the validity of local regulation of firearms, as well as relaxing the requirements for carrying a weapon in a vehicle. This bill removed the requirement for plain sight if the gun was holstered or in a locked container, and allowed carry in an unlocked but closed or latched container if in plain sight.

Oklahoma

When carrying a firearm with a concealed carry license, the handgun must be completely concealed. Under *21 OS § 1290.6* it is illegal to carry a handgun larger than .45 caliber pursuant to the Oklahoma Self-Defense Act (licensed concealed carry).

Oregon

In *Oregon*, the right to possess arms is protected by Article 1, Section 27 of the state constitution.

Oregon is a *shall-issue* concealed-carry state and is notable for having very few restrictions on where a concealed firearm maybe carried.[1] Oregon also has statewide *preemption* for its concealed-carry laws—meaning that, with limited exceptions, counties and cities cannot place limits on concealed-carry beyond those provided by state law.

Oregon is also an *open-carry* state, but preemption does not apply to open carry, so cities and counties are free to limit public possession of firearms by individuals who do not have a concealed carry permit. The cities of *Portland*, *Beaverton*, *Tigard*, *Oregon City*, *Salem*, and *Independence* have banned loaded firearms in all public places.

Pennsylvania

In a "city of the first class" (The number of residents equals 1 million or more. *Philadelphia* is the only such city in *Pennsylvania*), a license to carry a firearm (LTCF) is also needed to openly carry a firearm (unconcealed). There are multiple times when carrying a gun without a license is allowed like going hunting or to the firing range, but unless your activities fall under one of these exceptions, a license is required to carry a handgun. Nonetheless, all Pennsylvania LTCF permits are valid in Philadelphia.

Rhode Island

Although Rhode Island is a shall issue state by the 'local licensing authority' (*RI Gen Law 11-47-11*), they will revert you to the Attorney General when applying which is a *may* issue licensing authority (*RI Gen Law 11-47-18*). The local Police Chief has to sign ones application to 'verify residency' before one can submit the application to the attorney general. But, before he signs the application he may have you take an NRA Safety Course from a Certified NRA Instructor within the state. After one receives the training, the local BCI unit may also make you wait four to six weeks to get ones fingerprints on a FBI card needed for a new application. The whole process could take around three to four months before you can submit the application for consideration and then you will be told to allow up to ninety days for a response.

South Carolina

South Carolina is a "shall issue" concealed carry permit state. South Carolina also has "Castle Doctrine" legal protection of the use of deadly force against intruders into one's home, business, or car. Open carry is not allowed, but no permit is required to carry a loaded handgun in the console or glove compartment of a car. As of September 12, 2008, states with which South Carolina has reciprocity are: Alaska, Arizona, Arkansas, Florida, Kansas, Kentucky, Louisiana, Michigan, Missouri, North Carolina, Ohio, Texas, Tennessee, Virginia, West Virginia, and Wyoming.

South Carolina law also now supports a "stand your ground" philosophy under the "Protection of Persons and Property Act" *section* 16-11-440(C) with the following language. The act was apparently ruled nonretroactive in State v. Dickey.

A person who is not engaged in an unlawful activity and who is attacked in another place where he has a right to be, including, but not limited to, his place of business, has no duty to retreat and has the right to stand his ground and meet force with force, including deadly force, if he reasonably believes it is necessary to prevent death or great bodily injury to himself or another person or to prevent the commission of a violent crime as defined in Section 16-1-60.

South Dakota

South Dakota is a "shall issue" state for concealed carry. The local county sheriff shall issue a permit to carry a concealed pistol to qualified applicants. A temporary permit shall be issued within five days of the application. Concealed carry is not permitted at an elementary or secondary school, in a courthouse, or in any establishment that derives over half of its income from the sale of alcoholic beverages. For nonresidents, South Dakota recognizes valid concealed carry permits from any other state.

Open carry is legal in South Dakota and does not require a concealed pistol permit. Firearms maybe transported in vehicles if they are clearly visible.

When buying a handgun from a Federal Firearms License (FFL) holder, an application to purchase a handgun must be filled out by the buyer and submitted to the local police by the seller. Beginning June 1, 2009, anyone who passes the federal background check will be able to take possession of any firearm immediately, per SB0070.

South Dakota has state preemption of firearms laws. Units of local government may not restrict the possession, transportation, sale, transfer, ownership, manufacture, or repair of firearms or ammunition or their components.

Firearms manufacturers, distributors, and sellers are not liable for injury caused by the use of firearms.

Tennessee

Tennessee State Constitution, Article I, Section 26, reads:
That the people have the right to keep and possess arms for their common defense; but the Legislature shall have the power, by law, to regulate the wearing of arms with a view to prevent crime.

State Supreme Court rulings and state attorney general opinions interpret Section 26 to mean regulation cannot and should not interfere with the common lawful uses of firearms, including defense of the home and hunting, but should only be aimed at criminal behavior. "Going armed," carrying any sort of weapon for offense or defense in public, is a crime, except carrying a handgun for defense is allowed with a state-issued permit. At one time, Tennessee required a purchase permit for a handgun approved by one's city police chief or county sheriff with a fifteen-day waiting period; that was replaced under the federal Brady Act with the Tennessee Instant Check System (TICS).

Texas

Texas has no laws regarding possession of shotguns, or rifles by persons 18 years or older without felony convictions; 21 years or older for handguns. NFA weapons are also only subject to Federal restrictions; no State regulations exist. Municipal and county ordinances on possession and carry are generally overridden (*preempted*) due to the wording of the Texas Constitution, which gives the Texas Legislature (and it alone) the power to "regulate the wearing of arms, with a view to prevent crime." Local ordinances on discharge of a firearm are generally allowed under this preemption.

A rifle, shotgun, or other long-barreled firearm maybe carried openly, although there is debate as to whether doing so constitutes "disorderly conduct" (which defines an offense, in part, as "displaying a firearm or other deadly weapon in a public place in a manner calculated to cause alarm"). *Open carry* of a handgun in public is generally illegal in Texas; exceptions include when the carrier is on property he/she owns or has lawful control over, while hunting, or while participating in some gun-related public event such as a gun show. A permit to carry concealed is thus required to carry a handgun in public. Concealed carry permits are issued on a nondiscretionary basis ("shall-issue") to all qualified applicants. In addition, Texas recognizes most out-of-state concealed-carry permits.

The concealed handgun law sets out the eligibility criteria that must be met. For example, an applicant must be qualified to purchase a handgun under the state and federal laws. Additionally, a number of factors may make a person ineligible (temporarily or permanently) to obtain a license, including: felony convictions (permanent) and Class A or B misdemeanors (5 years, permanent in cases of domestic violence), including charges that resulted in probation or deferred adjudication, pending criminal charges (indefinite until resolved), chemical or alcohol dependency (defined as 2 convictions for substance-related offenses; 10-year ban from the date of the first conviction), certain types of psychological diagnoses (indefinite until the condition is testified by a medical professional as being in remission), protective or restraining orders (indefinite until rescinded), or defaults on taxes, governmental fees, student loans or child support (indefinite until resolved). This last category, though having little to do with a person's ability to own a firearm, is in keeping with Texas policy for any licensing; those who are delinquent or in default on State-regulated debts are generally barred from obtaining or renewing any State-issued license, as an incentive to settle those debts.

A person wishing to obtain a CHL must also take a State-set instruction course covering topics such as applicable laws, conflict resolution, and handgun safety, and pass a practical qualification at a firing range with a weapon of the type they wish to use (revolver or semi-automatic) and of a caliber greater than .32." They may then apply, providing a picture, fingerprints and other documentation, to the DPS, which processes the application, runs a background check, and if all is well, issues the permit.

On March 27, 2007, Governor *Rick Perry* signed Senate Bill 378 into law, making Texas a "Castle Doctrine" state which came into effect September 1, 2007. Residents lawfully occupying a dwelling may shoot a person who "unlawfully, and with force, enters or attempts to enter the dwelling," or who removes or attempts to remove someone from that dwelling, or who commits or attempts to commit a "qualifying" felony such as burglary, arson, rape, aggravated assault, robbery or murder. In addition, a shooter who has a legal right to be wherever he/she is at the time of a defensive shooting has no "duty to retreat" before being justified in shooting; the "trier of fact" may not consider whether the person retreated when deciding whether the person was justified in shooting.

Gov. Perry also signed HB 1815, a bill that allows any Texas resident to carry a concealed handgun in the resident's motor vehicle without a CHL or other permit. Chapter 46, Section 2 of the Penal Code states that it is in fact not "Unlawful Carry of a Weapon" for a person to carry a weapon while in a motor vehicle they own or control, or to carry while heading directly from the person's home to that car. However, lawful carry while in a vehicle requires these three critical qualifiers: (1) the gun must be concealed; (2) the carrier cannot be involved in criminal activities; and (3) the carrier cannot be a member of a criminal gang.

Possession of automatic firearms, short-barreled shotguns or rifles, or silencers is permitted, if the weapons have been federally registered in accordance with the *National Firearms Act*.

Utah

Utah allows for *open carry* of *unloaded* firearms without a concealed firearm permit. "Unloaded" as it applies here, means that there is no round in the firing position, and the firearm is at least two "mechanical actions" from firing. As carrying the firearm with the chamber empty, but with a full magazine, meets this definition (the handler must chamber a round, and then pull the trigger), this is a common work around for Utah residents who do not wish to acquire a permit. Without the permit, the firearm must be clearly visible. Utah requires a permit to carry a concealed firearm. With a permit, a person may carry a firearm with a loaded chamber either openly or concealed. *Utah will honor a permit issued by any state or county.*

Utah law allows for a "Non-Resident" Concealed Firearm Permits to be issued. The Utah Concealed Firearm Permit is valid in thirty-four states across the US. However there are several states that have passed statutes that do not honor a "Non-Resident" permit. For example, Colorado will honor Utah's permit, but the permittee must be a resident of Utah for his permit to be valid. Utah concealed firearm permits are "shall issue" and will be issued to anyone meeting the requirements.

Utah law recognizes a permit to carry a concealed firearm issued by any state or county (76-10-523(2)(b)).

Utah is a "Stand Your Ground" state, in which there is no duty to retreat before use of deadly force, the person reasonably believes that a perpetrator is going to commit a forcible felony in the habitation, and that the force is necessary to prevent the commission of the felony.

In Utah a person may carry firearms in many places not allowed by other states, including (but not limited to): banks, *bars*, and state parks. With a valid Utah concealed firearm permit you may also carry in schools (K-12 and public colleges). *Utah's Uniform Firearm Laws* expressly prohibits public schools from enacting or enforcing *any* rule pertaining to firearms. Accordingly, Utah is the only state in the Union that *requires* public schools to allow lawful firearms possession.

Utah weapon laws can be found at *the Utah State Legislature*

Vermont

Vermont has very few gun control laws. Gun dealers are required to keep a record of all handgun sales. It is illegal to carry a gun on school property or in a courthouse. State law *preempts* local governments from regulating the possession, ownership, transfer, carrying, registration, or licensing of firearms.

The term "Vermont Carry" is used by gun rights advocates to refer to allowing citizens to carry a firearm concealed or openly without any sort of permit requirement. Vermont law does not distinguish between residents and nonresidents of the state; both have the same right to carry while in Vermont.

The Vermont constitution of 1793, based partly on the U.S. Constitution and Bill of Rights, guarantees certain freedoms and rights to the citizens: "That the people have a right to bear arms for the defense of themselves and the State—and as standing armies in time of peace are dangerous to liberty, they ought not to be kept up; and that the military should be kept under strict subordination to and governed by the civil power."

Virginia

The right to keep and bear arms is protected by the Constitution of *Virginia*.

The Constitution of Virginia: Article I, Section 13. Militia; standing armies; military subordinate to civil power.

That a well regulated militia, composed of the body of the people, trained to arms, is the proper, natural, and safe defense of a free state, therefore, the right of the people to keep and bear arms shall not be infringed; that standing armies, in time of peace, should be avoided as dangerous to liberty; and that in all cases the military should be under strict subordination to, and governed by, the civil power.

There is State preemption of local firearm laws.

Localities may regulate:

- The discharge of firearms.
- The transportation of a loaded rifle or shotgun.
- Fingerprinting for concealed handgun permits, though fingerprints may not be kept and must be destroyed or returned to the applicant following the background check.
- The governing body of any county may require sellers of pistols and revolvers to furnish the clerk of the circuit court of the county, within ten days after sale of any such weapon, with the name and address of the purchaser, the date of purchase, and the number, make and caliber of the weapon sold.
- The use of pneumatic guns.

The following firearms are prohibited in Virginia:

- o "Sawed-off" shotgun or rifle (<18" smooth bore barrel, <16" rifled barrel). Overall length of a rifle must be >26".
- Striker 12, aka "streetsweeper," or any semi-automatic folding stock shotgun of like kind with a spring tension drum magazine capable of holding twelve shotgun shells.
- *NFA34* weapons are allowed under Virginia state law but *machine guns* must be registered with the *Virginia State Police*[227].

Firearms are prohibited from the following places:

- Places of religious worship, without good and sufficient reason.
- Any courthouse.
- An air carrier airport terminal.[230]
- Loaded firearms that hold more than 20 rounds, or a shotgun that holds more than 7 shells, on any public street, road, alley, sidewalk, public right-of-way, or in any public park or any other place of whatever nature that is open to the public in the Cities of Alexandria, Chesapeake, Fairfax, Falls Church, Newport News, Norfolk, Richmond, or Virginia Beach or in the Counties of Arlington, Fairfax, Henrico, Loudoun, or Prince William.

- The property of any public, private or religious elementary, middle or high school, including buildings and grounds; that portion of any property open to the public and then exclusively used for school-sponsored functions or extracurricular activities while such functions or activities are taking place; or any school bus owned or operated by any such school. Although a person with a Concealed Handgun Permit is allowed to have their weapon with them on school property so long as they remain inside a vehicle. Should they exit the vehicle, the firearm must remain inside the vehicle while on school property.
- Concealed handguns are prohibited from any restaurant or club that is licensed to sell alcohol for on-premises consumption. Openly carried handguns are permitted in such places.
- Private property where prohibited by the owner.[234]

The following persons are prohibited from possessing a firearm in Virginia:

- ... Acquitted by reason of insanity.
- ... Adjudicated legally incompetent, mentally incapacitated,
- ... Involuntarily admitted to a facility or ordered to mandatory outpatient treatment
- ... Subject to protective orders.
- ... Convicted of certain drug offenses; for a period of five years.
- ... Who have been convicted of a felony, kidnapping, robbery by the threat or presentation of firearms, or rape.
- ... Not a citizen of the United States or who is not a person lawfully admitted for permanent residence.
- ... Under the age of eighteen; while outside of his home or property without parental permission and adult supervision. A child over the age of twelve may use a firearm while unsupervised, in a place they have been granted permission by the property owner, if the child has been authorized by a parent, guardian, person standing in loco parentis to the child or a person twenty-one years or over who has the permission of the parent, guardian, or person standing in loco parentis to supervise the child in the use of a firearm.

Concealed firearms

Virginia *shall issue* a Concealed Handgun Permit (CHP) to any qualified person, 21 years of age or older who applies in writing to the clerk of the circuit court of the county or city in which he resides. Virginia also issues nonresident permits to qualifying individuals. The permit may cost no more than $50 for residents, and $100 for nonresidents. The permit is valid for five years, but can be revoked for unlawful activities. The CHP does not permit the carrying of any concealed weapons enumerated in § 18.2-308 except a handgun. A CHP holder may not carry a concealed weapon into a restaurant or club licensed to serve alcohol for on-premises consumption. A CHP holder, while carrying a concealed handgun, may not be under the influence of alcohol or illegal drugs. A conviction of driving while intoxicated [§ 18.2-266] or public intoxication [§ 18.2-388] are examples of prima facie evidence that the person is "under the influence." Virginia maintains concealed handgun permit reciprocity with other States and recognizes some licenses from other States without a formal reciprocity agreement. The list of such states is maintained by the Virginia State Police. The following persons are prohibited from applying for a concealed handgun permit: Any prohibited person enumerated above.

- Persons under the age of 21.
- An individual who has been convicted of two or more misdemeanors within the five-year period immediately preceding the application, if one of the misdemeanors was a Class 1 misdemeanor.
- An individual who has been convicted of any assault, assault and battery, sexual battery, discharging of a firearm in violation of § 18.2-280 or 18.2-286.1 or brandishing of a firearm in violation of § 18.2-282 within the three-year period immediately preceding the application.
- An individual who has been convicted of stalking.
- An individual who has received mental health treatment or substance abuse treatment in a residential setting within five years prior to the date of his application for a concealed handgun permit.
- An individual who is addicted to, or is an unlawful user or distributor of, marijuana or any controlled substance.

Virginia Concealed Handgun Permit (CHP) holders are exempt from:

- One handgun a month law
- Carrying a semi-automatic center-fire rifle or pistol loaded with 20 rounds or more in certain, prohibited, public areas.
- Gun Free School Zone act, CHP holders are allowed to have guns on school grounds in their personal vehicles as long as they stay in the car and the gun remains concealed
- General College Carry Restrictions
- Ban regarding firearms in VA General Assembly.

- Ban of firearms from State parks.

Purchasing of firearms

- A person may not sell or otherwise furnish firearms to any person he knows is prohibited from possessing or transporting a firearm pursuant to § 18.2-308.1:1, 18.2-308.2, subsection B of § 18.2-308.2:01, or § 18.2-308.7.
- Dealers must perform a criminal background check. Nonresidents may purchase long guns and handguns, but a handgun purchase requires a report from the Department of State Police.
- One handgun per 30-day period.]
  - Except:
    - With an enhanced background check.
    - Law enforcement, security companies,
    - A person whose handgun has been lost or stolen.
    - When a handgun is traded in at the same time as one is purchased.
    - A person who holds a valid CHP.
    - A person conducting private purchases.

Open Carry

- Legal when the firearm is not hidden from common observation.
- The minimum age to Open Carry (OC) is 18.
- It is more common in rural areas of Virginia.
- It is uncommon in urban areas, but not unheard of nor illegal.
  - Local law enforcement in some areas has been known to stop and question individuals openly carrying firearms.
- It is legal to openly carry in a vehicle.
- Since concealed carry is prohibited in restaurants and clubs that serve alcohol, it is common for someone carrying a concealed firearm to tuck their shirt behind the butt of their handgun when they enter the premises. This is known as the "Virginia Tuck."

Washington

Washington is one of the original "shall issue" states. The county sheriff or city police chief shall issue a concealed pistol license to any applicant, age 21 or older, who meets certain requirements, including no felony convictions, no misdemeanor domestic violence convictions, and no outstanding warrants.

Open carrying of firearms is not prohibited by law although trouble with some law enforcement agencies has been encountered while open carrying in the past, most notably in a case in *Ellensburg, Washington.*

Currently, there is a growing movement toward open carry in Washington. In Washington, there was a tremendous amount of disinformation among law enforcement officers, gun store employees, and firearms instructors about *RCW 9.41.270.*

In December 2005, activists Lonnie Wilson and Jim March went to the state archives in Olympia to research the origins of the law. March, with his experience in researching gun control laws created out of racial discrimination and strife in California, surmised during a conversation between himself and Wilson that due to the year it was passed, it was likely due to "Panther paranoia." March was proven correct.

The law, passed in 1969, was passed in response to incidents involving the Seattle Chapter of Black Panther Party at *Rainier Beach High School* and the *Protest of the Mulford Act* by the main organization in the California Assembly.

Due to the fact that Washington State Constitution contains an individual right-to-keep-and-bear-arms provision (Article 1, Section 24), the Washington Legislature revised the bill that was debated to remove the "within 500 feet [152 m] of a public place" provisions and left the current statute as is. There were points of debate about whether this could be interpreted as an open-carry ban, to which the sponsors of the bill replied that it was a ban against the type of intimidation that the *Black Panther Party* engaged in at Rainier Beach and the California Assembly, not an open-carry ban.

Many law enforcement personnel, a generation removed from the events and discussions of the Legislature when the law was created and without much guidance, interpreted the law passed as an open-carry ban that is situational to someone making a 911 phone call. This interpretation spread to gun store employees and firearms instructors, who have a lot of personal interaction with law enforcement.

Using the information from the state archives, Wilson pursued the issuance of guidance and memorandums to individual officers by police administrators. After one department issued one of these bulletins, Wilson acquired the bulletin by a public-records request, used the training bulletin as a template, and approached most police departments throughout the state. To this day, over a dozen major departments, including the King County Sheriff's Department and the Seattle Police Department, have issued advisories and roll call training to their officers that peaceable open carry of a handgun in a holster is legal.

As a general rule, a person may legally open-carry in Washington State in any place it is legal to possess a loaded handgun. To open-carry in a vehicle (i.e., car, bus, etc . . .) a person must have a valid concealed pistol license. Some police agencies can be unfriendly toward open-carry, so it is important that before a person exercises their right to bear arms in this fashion they acquaint themselves with relevant laws.

Prohibited areas for firearms are contained in *RCW 9.41.300*, *RCW 9.41.280*, and *RCW 70.108.150*.

Per *RCW 9.41.290* (state preemption of firearm laws), divisions of local government (city, county, town, or other municipality) cannot regulate firearms more restrictively than the state does. Exceptions to state preemption—that is, areas in which local governments are allowed to regulate firearms—are contained in *RCW 9.41.300*. These exceptions include:

- "Restricting the discharge of firearms in any portion of their respective jurisdictions where there is a reasonable likelihood that humans, domestic animals, or property will be jeopardized. Such laws and ordinances shall not abridge the right of the individual guaranteed by Article I, section 24 of the state Constitution to bear arms in defense of self or others."
- "Restricting the possession of firearms in any stadium or convention center, operated by a city, town, county, except that such restrictions shall not apply to [concealed pistol license holders, law enforcement officers, or any] showing, demonstration, or lecture involving the exhibition of firearms."
- "Restricting the areas in their respective jurisdictions in which firearms maybe sold."

Several localities (including transit agencies) who had wrongfully enforced preempted local ordinances and rules have been challenged by activists in the open-carry movement (who are most directly affected by the enforcement of such ordinances) and have since backed down from enforcement and directed their police departments no longer to enforce the ordinances and rules.

Washington allows ownership of a firearm silencer, but using one is prohibited by *RCW 9.41.250(1c)* which makes it a gross misdemeanor to "Use any contrivance or device for suppressing the noise of any firearm."

Washington is a "Stand Your Ground" state, in which there is no duty to retreat in the face of what would be perceived by an ordinary person to be a threat to themselves or others by another person that is likely to cause serious injury or death.

It is a Class C felony for a noncitizen to possess a firearm in Washington without an Alien Firearm License. [264] Washington had stopped issuing Alien Firearm Licenses due to a problem obtaining background checks, but a court ordered the Washington State Department of Licensing to resuming issuing the licenses after a lawsuit was filed by the *National Rifle Association* and *Second Amendment Foundation* on *Second Amendment* grounds.[265]

It is a gross misdemeanor to aim a firearm "whether loaded or not, at or toward any human being."

Washington State accepts concealed-weapons permits from the following states: Louisiana, Michigan, Mississippi, North Carolina, Ohio, Oklahoma, and Utah. Washington State law also carves exemptions into state law regarding Concealed Pistol Licenses. Perhaps the most interesting is RCW 9.41.060, section 8: "Any person engaging in a lawful outdoor recreational activity such as hunting, fishing, camping, hiking, or horseback riding, only if, considering all of the attendant circumstances, including but not limited to whether the person has a valid hunting or fishing license, it is reasonable to conclude that the person is participating in lawful outdoor activities or is traveling to or from a legitimate outdoor recreation area;"[268]. This little known law essentially allows vehicle and concealed carry WITHOUT a CPL as normally required in 9.41.050 as long as you meet the provisions of that section.

West Virginia

In West Virginia permits aren't required to possess handguns. A permit test must be passed and a license acquired to carry a concealed handgun. West Virginia also allows open carry. State laws are *preempted*, but there still remain *grandfathered* restrictions on open carry in some localities, such as *Charleston*[269] and *Dunbar*. The *SB 716* is a proposed senate bill, which would amend the law to remove those restrictions.

West Virginia enacted the *castle doctrine* on *April 10, 2008*.

Wisconsin

Wisconsin is one of two states that completely prohibit concealed carry by private citizens. Open carry is legal except where prohibited by law (government buildings, schools, and establishments that sell liquor) but, in the words of one Wisconsin resident, "you will attract the attention of every police officer in the area." Some jurisdictions have tried to prosecute open-carry by equating the open carry of handguns with disorderly conduct. Inside vehicles, the firearm must be both unloaded and encased; having a loaded firearm on the front seat was held to be concealed and therefore illegal in a 1994 case. Bills to enact "shall-issue" were twice vetoed by Governor *Jim Doyle* in January 2004 and again in January 2005 after passing in both houses of the Wisconsin legislature. In 2005, the Assembly fell two votes short of overriding Doyle's veto.

Other laws

Possession of a firearm while intoxicated, shooting within 100 yards (91 m) of a home without permission, pointing a weapon at anyone except in self-defense, and negligent handling of a weapon are all outlawed. Statute 941.20

Carrying a concealed weapon is a class A misdemeanor, state statute 941.23. This is any "weapon," not just firearms. Knives are legally defined as "dangerous weapons."

Going armed in any building owned/leased by the government is a class A misdemeanor, state statute 941.235.

Carrying a handgun where alcohol is sold/consumed is generally a class A misdemeanor, state statute 941.237.

Armor-piercing ammunition prohibited in handguns when committing a crime. Statute 941.296.

"No person may carry or display a facsimile firearm in a manner that could reasonably be expected to alarm, intimidate, threaten or terrify another person," unless on your own property or business, or that of another person with their consent. Statute 941.2965.

Wisconsin has a state preemption law that generally forbids cities from passing firearms ordinances more strict than that of state law. Statute 66.0409. This doesn't affect zoning regulations, which is why only one *Madison* gun shop sold handguns. That shop, along with others as well as the only gun club in *Middleton*, have closed.

Committing a crime while possessing a dangerous weapon is a penalty enhancer. Statute 939.63.

It is a felony to possess a firearm if you:

- Have been convicted of a felony
- Committed a felony as a juvenile
- Have been found not guilty of a felony by reason of mental disease or defect
- Have been committed under mental health laws and ordered not to possess a firearm
- Are the subject of a domestic-abuse or child-abuse restraining order
- Are ordered not to possess firearms as a subject of a harassment restraining order.

Any person who knowingly provides a firearm to an ineligible person is party to a felony crime. Statute 941.29

Buying and selling

There is a 48-hour waiting period on handgun purchases from an FFL dealer (does not apply to private sales): Statute 175.35

Rifles and shotguns can be purchased in a contiguous state as long as the purchase complies with Federal law and the laws of the contiguous state. Statute 175.30

===State parks and wildlife refuges Statute 29.089 requires firearms to be unloaded and encased in state parks. There is an exception for hunting when the hunt is administratively approved. Statute 29.091 requires firearms to be encased and unloaded in wildlife refuges.

Class 3 firearms

Machine guns are legal if you follow BATFE process, state statute 941.27

Short-barrel rifles and shotguns are legal if you follow BATFE process, state statute 941.28

Silencers are legal if you follow BATFE process, statute 941.298

Firearms and minors

It is a class I felony to possess a firearm on school grounds or within 1000' of a school zone. Statute 948.605. This statute does not apply to:

- Private property not part of school grounds
- Individuals licensed by the local government body to possess the firearm
- Unloaded and encased firearms
- Individuals with firearms for use in a school-approved program
- Individuals with school contract to possess firearm
- Law enforcement acting in official capacity
- Unloaded firearms when traversing school grounds to gain access to hunting land, if the entry is approved by the school.

It is a class G felony to discharge or attempt to discharge a firearm in a school zone. Limited exceptions for private property not part of school grounds, school programs, and law enforcement.

Leaving a firearm within reach of a child under 14 is generally a misdemeanor, if that child points it at anyone or shows it to anyone in a public place. Defenses include having the gun locked in a safe or container, or having a trigger lock on the gun, or removal of a key operating part, or illegal entry by anyone to obtain the firearm, or a reasonable belief a juvenile couldn't access the firearm. Statute 948.55

Firearms retailers are required to provide every buyer with a written warning stating, "If you leave a loaded firearm within the reach or easy access of a child, you may be fined or imprisoned, or both if the child improperly discharges, possesses or exhibits the firearm." Statute 175.37

Possession of a dangerous weapon by anyone under 18 is a class A misdemeanor. Giving/loaning/selling a dangerous weapon to someone under 18 is a class I felony. Statute 948.60. Defenses to prosecution under this statute:

- Target practice under the supervision of an adult
- Members of armed forces under 18 in the line of duty.

For hunting purposes, the following exceptions to the age limit apply, as specified in statute 29.304 for weapons with barrels 12" or longer.

- Under 12 may not hunt with a firearm or bow under any circumstances
- Under 12 can only possess firearm/bow in Hunter Safety class, or while cased/unloaded and under parental supervision while going to/from Hunter Safety class
- 12 to 13 may hunt when accompanied by an adult
- 12 to 13 may possess firearm when accompanied by an adult, or while transporting cased/unloaded firearm to/from Hunter Safety class, or in Hunter Safety class
- 14 to 15 is the same as 12 to 13, except Hunter Safety graduates can hunt and possess firearms without adult supervision.

School students shall be suspended until their expulsion hearing if they possess a firearm in school or during a school event. State law requires a minimum one-year expulsion for this offense. Statute 120.13(1)(bm) and 120.13(1)(c)2m. In addition, the student's driver license maybe suspended for two years under Statute 938.34(14q). This suspension also applies to bomb threats and CCW violations in government buildings.

School Zones

Any individual who knowingly possesses a firearm at a place that the individual knows, or has reasonable cause to believe, is a school zone is guilty of a Class I felony. "School" means a public, parochial or private school which provides an educational program for one or more grades between grades 1 and 12 and which is commonly known as an elementary school, middle school, junior high school, senior high school or high school. "School zone" means any of the following: 1. In or on the grounds of a school. 2. Within 1,000 feet from the grounds of a school.

Firearms in vehicles

When openly carrying (the only legal method in Wisconsin) a firearm must be unloaded, cased, and put out of reach when in a vehicle. Because of this statute firearms must be frequently and unnecessarily handed, loaded, and unloaded to comply with the state's laws. In this section, unloaded = "Having no shell or cartridge in the chamber of a firearm or in the magazine attached to a firearm." Encased = "enclosed in a case that is expressly made for the purpose of containing a firearm and that is completely zipped, snapped, buckled, tied or otherwise fastened with no part of the firearm exposed." Statute 167.31

Boats: Firearms must be unloaded and encased when the motor is running.

Aircraft: Firearms must be unloaded and encased.

Cars, trucks, motorcycles, ATV, snowmobiles: Firearms cannot be placed in or on a vehicle unless the firearms are unloaded and encased. However, it is legal to "lean an unloaded firearm against a vehicle." Statute 167.31(4)(d).

Exceptions: Law enforcement officers, military personnel on active duty, landowners and their family and employees on farm tractors inside *CWD* eradication zones, and disabled hunters with special permits meeting all the requirements.

Wyoming

According to the office of the *Attorney General of Wyoming*, Wyoming state law (WS § 6-8-104) provides for the issuance of concealed firearm permits. Wyoming also recognizes concealed firearms permits from the following list of other states in the United States (subject to frequent review and revision): Alabama, Alaska, Arizona, Colorado, Florida, Georgia, Idaho, Indiana, Kentucky, Louisiana, Michigan, Mississippi, Montana, New Hampshire, New Mexico, Ohio, Oklahoma, Pennsylvania, South Carolina, South Dakota, Tennessee, Texas, and Utah.

Concealed carry in the United States
From Wikipedia, the free encyclopedia

In the *United States, carrying a concealed weapon* (*CCW*, also known as *concealed carry*) is the legal authorization for private citizens to carry a *handgun* or other weapons in public in a concealed manner, either on the person or in close proximity to the person. In some states, it is sufficient to be a resident or permanent resident (*greencard* holder). Under current federal legal precedent, it is considered "constitutional" under the *Second Amendment* for states to have concealed carry licensing that permits concealed carried weapons, or even not to require any permits for concealed carry weapons; for example, any legal gun owner in the state of Vermont may carry concealed weapons with no permitting required. It is likewise "constitutional" under the Second Amendment for states to have laws that prohibit concealed carried weapons, although only two states have done so. Laws governing concealed carried weapons vary from state to state. Some states restrict concealed carried weapons to a single handgun, whereas others permit multiple handguns or martial arts weapons to be carried.

Various states give different terms for licenses or permits to carry a concealed firearm, such as a *Concealed Handgun License/Permit* (CHL/CHP), *Concealed (Defensive/Deadly) Weapon Permit/License* (CDWL/CWP/CWL), *Concealed Carry Permit/License* (CCP/CCL), *License To Carry (Firearms)* (LTC/LTCF), *Carry of Concealed Deadly Weapon license* (CCDW), and similar. Many states that issue licenses to carry a concealed handgun also allow the practice of *open carry* by the license holder, and another 31 states allow open carry without any license.

Although the current trend toward adopting concealed carry laws has been met with *opposition*, no state which has adopted a "Shall-Issue" concealed carry law has reversed its decision. As of February 2008[update], 48 US states allow some form of concealed carry[5] (though nine of them have discretionary "may-issue" policies, a few of these being effectively "no-issue" in practice) and all but six provide for some variant on nonconcealed "open-carry." The states of Wisconsin and Illinois, and the District of Columbia do not have any form of concealed-carry licensing; Wisconsin allows for *open carry* in most situations, while Illinois only allows it in rural areas subject to county restriction.

On March 19, 2009, a federal judge ordered a temporary restraining order blocking the implementation of the rule allowing concealed carry permit holders to carry firearms concealed within National Park Service lands within states where their permits are valid, based upon environmental concerns, in response to concerns by the Brady Campaign to Prevent Gun Violence, the National Parks Conservation Association, and the Coalition of National Park

Service Retirees.[78] On May 22, 2009 President Obama signed a law (HR 627) that will prohibit the Secretary of the Interior from enacting or enforcing any regulations that restrict possession of firearms in National Parks or Wildlife Refuges, as long as the person complies with laws of the state in which the unit is found.

You have just examined the state-by-state gun laws. What do they show? They show that you have to meet certain requirements and statues before you can obtain, carry and use a firearm and ammunition, however once you meet those requirements, what you do with your gun is entirely up to you without no accountability of what you are doing with deadly bullets that kill. A person can literally become just like the Washington sniper of the 1990s and just randomly go out and shoot and kill folks because they are mad at the world. And do this without any accountability as to what they are doing with deadly bullets that kill. And since this person is not being challenged by the current state by state gun laws as to what he or she are doing with those deadly bullets. No one will know who the sniper is that's doing the killing unless he or she is caught. By then it's too late. Look at how many innocent lives was lost due to failure of the current gun laws to challenge and prevent repeated offenses.

The current state by state gun control laws are not setup to prevent they are setup to restrict and enforce. We need a gun control law that can prevent repeated gun violence and homicides. Just like you read in the story in Chapter 2 of this book. Tracking the bullet is the only law that can be passed that will do this without violating anyone's right to bear arms under the second amendment.

Eventually what you will start to see happening is that the gun crimes will start to go down because the bad guys are being exposed, because of being challenged as to what they are doing with deadly bullets that they purchased. And they will be taken out of the bullet purchasing system forever. This fact alone will save hundreds of lives.

Young gangbangers will lose their suppliers because their suppliers are being eliminated from purchasing deadly bullets that kill because of their irresponsible acts of illegally supplying the criminal element with deadly bullets that kill. You will see more gifted kids grow up and become something great because their lives are not being cut short due to gun-related homicides.

People will start to feel safer now that the gun crimes are declining because of this new law called tracking the bullet to save lives. There will be less gun-related accidents and incidents and less people going to the hospital because of gun-related incidents due to negligence and irresponsible use of bullets that injure and kill.

Make this a law and start to save lives and create jobs and stimulate the economy before the out of control gun violence gets worse than what it is now and claim your child's life or maybe even yours at the pull of a trigger in the hands of an irresponsible one.

One thing is for certain and the statistics show this. If we do not work hard to pass a gun homicide prevention law called tracking the bullet to save lives. There is nothing in the current state-by-state gun laws that can prevent repeated gun offenses done by this new breed of young teenage shooters. The current gun laws can only restrict to a certain degree. Gun homicides will continue to go up not down and the statistics show this to be true. The time to act is now by pushing the state lawmakers and federal lawmakers to pass this bill. Therein lies the common sense to an out-of-control problem.

Chapter 7

Restrict and Enforce Versus Challenge and Prevent

While I agree that there should be gun-control laws that set restrictions on obtaining, carrying, and using firearms and ammunition and then enforce the penalties on those who violate these imposed restrictions, I also realize that much more is needed than just setting restrictions. Do you think the criminal mind cares whether he legally obtains a license to carry a concealed weapon or not. Or cares about obtaining a firearms identification card to purchase guns and ammunition? The only thing that's on their minds is where can I get a loaded gun from to commit a crime. And there are many guns out on the streets today so that is a mute point. What about the sneaky criminal character who meets the required restrictions and statues to obtains a gun and then goes out there secretly committing gun violence and homicides. The Tracking the Bullet gun regulating system is designed to catch individuals like this and deactivate their rights. It is also designed to catch those who are providing bullets to the criminal street thugs and gangbangers by challenging what they are doing with the bullets that they have purchased. It will deny them the right to ever purchase bullets again.

Do guns kill people? The obvious answer to this question is guns can kill people but not alone. There has to be more pieces added to the equation in order to complete a gun homicide.

With a gun sitting on the table with nobody around that gun is no more lethal than a butter knife in the kitchen drawer. Now let's say we put bullets in that gun, it's still no more lethal than the butter knife in the kitchen drawer. Now let's add a person to the equation who picks up that loaded gun and while we're at it let's add the characteristics of this person.

Let's say that this person is an adolescent. What about his environment? He lives in a poverty stricken neighborhood with very little education if any.

He comes from a single parent household and necessity dictates that his parent works during the day, and there is no father figure or role model around.

His substitute family, since the single parent is at work most of the day, are the gangbangers who shoot and kill people. He is around violence 24/7, drugs, rape, and gun violence in the streets is the order of the day.

There are drug dealers and buyers on every corner, liquor stores on every corner and outside the liquor stores are drunks constantly cursing and fighting each other all day long, drive-by shootings and gang initiations are also pretty common in his environment. He himself has been beaten up and robbed so many times that he lost count and he feels that this is normal behavior for anybody that lives in this type of environment.

There is constant murder and gang-related activity around him. Anger and hatred is on his mind constantly because of the environment that he lives around every day.

He has not been taught to respect himself or nobody else for that matter. He could care less whether he lives or dies from one day to another and he feels that way about everybody else.

And now he is holding that loaded gun in his hand. That gun now becomes more lethal than hazardous goods traveling on a freight train.

Now multiply this type of scenario several hundred times and that's what you have in major metropolitan areas in the United States alone.

Common sense would dictate that we come up with a solution to keep deadly bullets from ever being loaded into this gun thereby making it less lethal in the hands of this type of character.

To be fair, The characteristics of a person could be any Individual not necessarily poverty stricken. What matters the most is the type of character holding that loaded gun.

Remember the Columbine high school shootings, although the perpetrator's environment may not have been poverty stricken as mentioned above. However, the individual that shot and killed and injured several individuals in the columbine shootings was not prevented from getting his hands on bullets loaded into a gun that killed innocent students.

Common sense would also dictate that the current gun-control laws that are on the federal books today are not all that is needed to combat the current gun homicides that are happening all over this country by young people.

If they were then you would have a decline in gun homicides and not an increase in gun homicides. We live in a different time than the 1930s, 1940s, or 1950s, where respect for authority and laws have largely diminished since then.

So why can't all of these restrictive gun control laws and statues stop these young individuals from shooting and killing people?

Because these laws and statues are largely based on an assumption that people will abide by these laws or else face the consequences or face the penalties for violating these laws and statues if they are caught.

More is needed to help prevent these young murderers from getting their hands on deadly bullets that kill. We need to attack the problem from the source. What gun control law can challenge the way a person is using his gun responsibly or not?

The answer is obvious. It's time to make Tracking the Bullet a state and federal law.